# THE LITTLE GREEN BOOK

Produced by Vole for The Green Alliance

Edited by

**Richard Boston,**
**Richard Holme**
**and Richard North**

Produced with more than a little help from

**John Adams, Michael Allaby, Maurice Ash, Gillian Darley, Robin Grove-White, Nigel Haigh, Christopher Hall, Gerard Morgan-Grenville, Barty Phillips, Hilary Senior, Harford Thomas, Colin Ward.**

Drawings and design by **Bryan Reading**

Published by Wildwood House, London

# PLANET EARTH

THE observable universe contains about one hundred thousand million galaxies. One of these is the Milky Way, which contains about one hundred thousand million stars. One of these stars is our Sun.

The Sun has nine planets, of which Earth is the fifth biggest. It is the only planet — indeed, the only known place in the universe — capable of producing and sustaining life.

Earth is thought to be 4,600 million years old. It takes the form of a slightly squashed sphere, since its polar axis (6,356,774 metres) is less than its equatorial axis (6,378,160 metres). Its polar circumference is 24,860 miles: its equatorial circumference is 25,326 miles.

Earth is enclosed by a bubble of air which consists of approximately 21 per cent oxygen, 78 per cent nitrogen, and .03 per cent carbon dioxide. Plants consume carbon dioxide and produce oxygen. Animals consume oxygen and give out carbon dioxide.

The surface of Earth is some 200 million square miles. About 70 per cent of this is sea, and 30 per cent land. Most of the land is in the northern hemisphere — 38 million square miles, against the southern hemisphere's 18,700,000 square miles.

The highest land is 29,028 feet above sea level (Mount Everest).
The deepest ocean bed is 36,198 feet at the Mariana Trench.
The hottest place on Earth is Dallol, Ethiopa (annual mean temperature 94°F 34°C)
The coldest is Polus Nedost Upnosti (annual mean temperature −72°F −57.8°C)
Highest recorded shade temperature, 136.4°F, 57.7°C, Al 'Aziziyah, Libya 13 September 1922.
Lowest ocean temperature, −126.9°F −88.3°C Vostok, Antartica 24 August 1960.
Strongest wind 231mph Mount Washington 1934.

| ERAS | | PERIODS AND EPOCHS | YEARS AGO | MAIN CHANGES |
|---|---|---|---|---|
| *PRECAMBRIAN ERA* | | | 4,600 Million | Volcanic eruptions Important ores formed |
| *PALEOZOIC Era of Ancient Life* | | Cambrian | 520 M | Age of vertebrates |
| | | Ordovician | 500 M | |
| | | Silurian | 440 M | |
| | | Devonian | 400 M | Age of fish |
| | | Carboniferous | 345 M | Coal deposited |
| | | Permian | 280 M | Larger amphibians |
| *MESOZOIC Era of Reptiles* | | Triassic | 230 M | Small dinosaurs |
| | | Jurassic | 180 M | Birds & Larger dinosaurs |
| | | Cretacious | 136 M | Crustaceans abound |
| *CENOZOIC Era of Mammals* | *Tertiary* | Eocene | 65 M | Mammals and Molluscs thrive. Ape like ancestors of man evolve |
| | | Oligocene | 40 M | |
| | | Miocene | 25 M | |
| | | Pliocene | 13 M | |
| | *Quaternary* | Pleistocene (Ice age) | 3 M | Glaciers spread, More varied mammals. Man evolves |
| | | Holocene | 8,000 to 10,000 | Glaciers melt. Warmer climate Larger land mammals die out. |

# PLANET EARTH

THE DINOSAURS are often quoted as a failure to adapt to a changing environment. They are dismissed as an evolutionary freak that Nature toyed with for a while before throwing it on the scrapheap and moving on to more sophisticated forms of life such as ourselves.

In fact the dinosaurs survived with remarkable success for 150 million years. By contrast Man is a very recent evolutionary development that has so far been going for only about 100,000 years. If we do as well as the dinosaurs did then we will have done extremely well.

The time span in which life on Earth has evolved is so great that most of us can only begin to understand it when it is scaled down. David Attenborough produced a comprehensible model in his BBC TV series *Life on Earth*, where he suggested that we should think of the history of life on this planet as lasting for one year so far, and that each day of that year is taken as representing 10,000,000 (ten million) years. On that scale we find that pioneering bacteria-like organisms (which the evidence of micro-fossils tells us existed 3,000 million years ago) appeared on or shortly after January 1. Dinosaurs were flourishing in the last few shopping-days before Christmas. Man arrived at about a quarter to midnight on New Year's Eve.

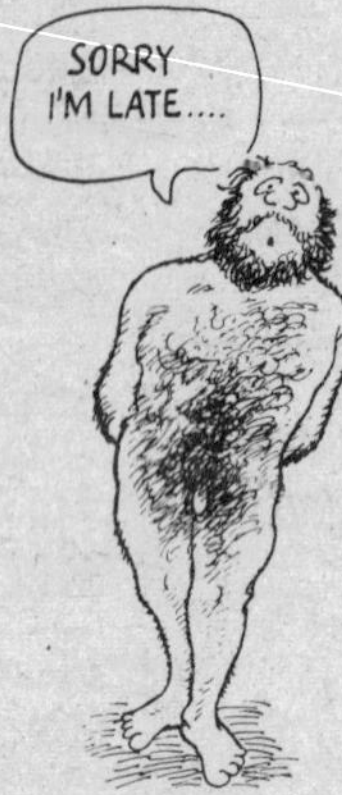
SORRY
I'M LATE....

**Man is Nature's sole mistake.**
*W. S. Gilbert* **(Princess Ida)**

# MAN

**What a piece of work is a man! How noble in reason! how infinite in faculty! in form, in moving, how express and admirable! in action how like an angel! in apprehension how like a god! the beauty of the world! the paragon of animals!**
*Hamlet*

WHEN we talk about "breeding like rabbits" we are either insulting rabbits or under-estimating our own abilities. At the very least, the pot is calling the kettle black.

Of all the people who have ever lived on earth, between 5 and 10 per cent are alive now.

More than one-third of the world's present population is aged less than 15.

The world's most crowded continent is Europe. Probably the most crowded countries in the world are Britain and the Netherlands. Britain has 229 people per square kilometre. England has 356 people per square kilometre and is four times as crowded as China, and 19 times as crowded as Russia.

The population of Neolithic Britain (2500-1900BC) was about 20,000. In 1750 it was 6 million. It has now settled down at around 56 million.

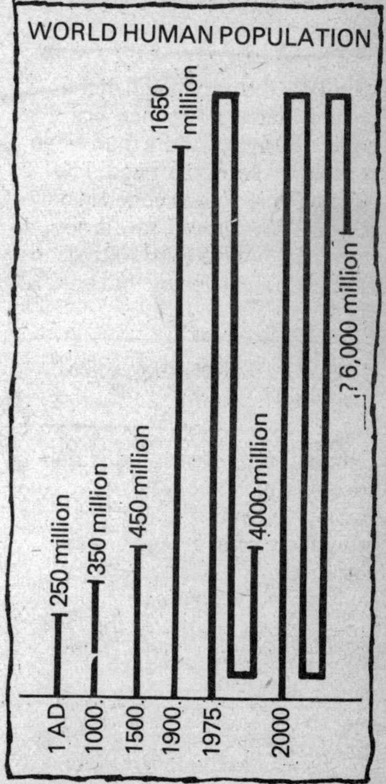
WORLD HUMAN POPULATION
250 million
350 million
450 million
1650 million
4000 million
?6,000 million
1 AD
1000
1500
1900
1975
2000

# MAN

**Man is a noble animal, splendid in ashes, and pompous in the grave. Sir Thomas Browne** *Urn Burial*

---

In 1917 the population of the United States of America was about 100 million more than when the Pilgrim Fathers arrived. The next 100 million was added in only 50 years. The growth rate is now declining, but the third 100 million should be achieved by the year 2000.

---

**Man is a tool-making animal.**
*Benjamin Franklin*

---

Industrial (and post-industrial) societies now mostly have stable population figures. The growth rates have more or less flattened out in Scandinavia, Switzerland, Belgium, Luxembourg, Germany, the United Kingdom and the United States of America.

---

**Man is the measure of all things.**
*Protagoras* **(as quoted by Plato)**

The developing world is growing as fast as the industrialised nations did when they were going through the same process.

650 million people live in India — twice its population 50 years ago. In 1800 Egypt's population was 2 million. Now it is 38 million.

---

**Glory to man in the highest! for Man is the master of things.**
*Algernon Charles Swinburne*

---

**The rich get richer . . .** Draw a line 40 degrees North of the Equator on our crowded planet. One-third of the world's population lives above this line. They have 5/6 of the Gross World Product. Two-thirds of the world's population live below this line. They have one-sixth of the Gross World Product.

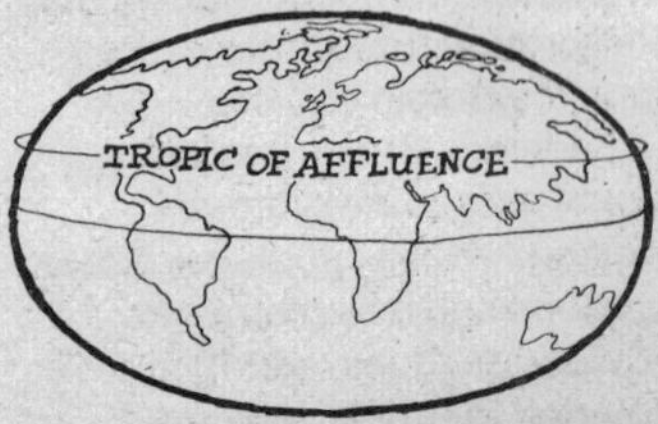
TROPIC OF AFFLUENCE

# TREES

**WITHOUT plants there would be no life on earth. Everything that breathes must have oxygen. The air contains about 21 per cent oxygen which we inhale at the rate of about 16 times a minute. We exhale about 16 per cent oxygen, using the other 5 per cent. Plants on the other hand absorb CO2 and give out oxygen. By means of photosynthesis the plant turns the sun's energy plus water plus CO2 into oxygen.**

ONE third of the world's trees grow in the Amazon's five million square kilometres (an area larger than Europe).

According to some scientists the Amazon forest produces through photosynthesis nearly half the world's oxygen. About 40 per cent of Amazonia's rain forests have disappeared in the past century, and at the present rate there will be precious little left by the end of the century.

The rain-forests of South-East Asia are also being cleared at a rate which, if it continues, will mean that there will be none left by the end of the country. Where replanting takes place it is usually with fast-growing temperate climate trees, which photosynthesize (and thus produce oxygen) at one tenth of the rate of tropical rain forest trees.

These massive tree clearings upset the water cycle, with effects that can lead in the short term to flooding and in the long-term to the creation of deserts. As the forests shrink, the deserts grow.

Trees make up more than three-quarters of the world's living matter. A threat to the trees is a threat to life.

Britain has fewer trees per square mile than any other European country except Iceland and Ireland. Barely 8 per cent of our land is wooded.

50 million trees have been lost in Britain in the last 25 years. 11 million elms have had to be felled

**continued over**

# TREES

**continued**

because of Dutch Elm disease. We need 100 million extra trees over the next 25 years to replace our losses.

*Forests and other wooded land as a percentage of total land*

| | |
|---|---|
| European Economic community | |
| Belgium | 20 |
| Denmark | 12 |
| France | 25 |
| German Federal Republic | 30 |
| Ireland | 4 |
| Italy | 27 |
| Luxembourg | 32 |
| Netherlands | 10 |
| United Kingdom | 8 |
| Scandinavian Countries | |
| Finland | 74 |
| Norway | 29 |
| Sweden | 64 |

The highest tree in the world is a 368-foot tall redwood (Sequoia sempervivens) in Redwood Grove, Humboldt County, California. The tree with the most massive trunk is the Montezuma cypress (Taxodium mucronatum) in Santa Maria del Tule, Oaxaca, Mexico. The circumference of the trunk is 160 feet, the diameter more than 50 feet.

---

**"On Sunday, December 3, 1961, the New York Times published a record edition in terms of bulk.**

**Each issue contained 678 pages and weighed 6 pounds 3¼ ounces. The press run was 1,458.558 copies.**

**Printing this staggering pile required 4,550 tons of paper, and newsprint, as everyone knows, is made from woodpulp.**

**Manufacturing the pulp for this single edition required the wood from over 77,000 full-grown trees!**

**The average age of these trees was seventy years, and cutting them left a 360-acre hole in the forest. It took all this to publish a single edition of a single newspaper in a single city, read in the morning and thrown away in the afternoon."**

**Andreas Feininger, Trees (Penguin)**

**Hedgerows:** Over the past 20 years, hundreds of thousands of miles of hedgerow have been removed by farmers wishing to make their fields bigger, to increase productivity. The rate of loss has been as follows:

| | | |
|---|---|---|
| 1945 — 1960 | 4,500 | miles per annum |
| 1960 — 1966 | 10,000 | " " |
| 1966 — 1970 | 4,500 | " " |
| 1970 — 1978 | 2,000 — 3,000 | " " |

Some Addresses

**The Tree Council** is an independent body set up in 1974 after the Department of the Environment's "Plant a tree in '73" campaign. Its function is to bring together all the national bodies which have a specific interest in trees.

**Woodland Trust**, Butterbrook, Harford, Ivybridge, Devon, was founded in 1972 with the aim of planting and safeguarding trees, and purchasing and maintaining endangered woods.

**Council for the Protection of Rural England**, 4 Hobart Place, London SW1.

**Men of the Trees**, Crawley Down, Crawley, Sussex.

**'For every old tree we enjoy we should plant ten new ones in gratitude to our ancestors and goodwill to our descendants. If the generations who follow us don't want them they can cut them down — there is nothing simpler, alas, than to fell in five minutes a tree which took fifty years to grow, but no way except by planting now to have trees for the twenty-first century.'**
**Nan Fairbrother**

20 hectares of tropical forest (about fifty acres) are cut down every minute of every day.
11 million hectares of trees disappear every year. 40 per cent of the tropical rain forest has already disappeared.

**continued over**

# TREES

**continued**

Scientists warn that continued massive burning of fossil fuels and levelling of forests will, by increasing the carbon dioxide content of the atmosphere, and disrupting the heat balance, eventually cause global temperature to rise and climates to change. Climatic shifts could in turn wipe out many species whose lives are closely attuned to particular environmental conditions. Ozone could also seriously damage the ecosystems.

The biggest single cause of extinctions over the next few years will be the destruction to habitats by growth of human populations and the encroachment by buildings into the natural environment. This will be particularly serious in tropical regions where the major species losses are predicted.

In the place where the tree falleth, there it shall be. **Ecclesiastes**

A fool sees not the same tree that a wise man sees. **William Blake**

---

# ANIMALS and PLANTS

- One in ten of the estimated 250,000 different plant species are said to be rare or threatened. These tend to be concentrated in the most vulnerable and species-rich habitats: islands, tropical moist forests, arid lands, Mediterranean ecosystems, wetlands and coastal sites.

A large number of plants are likely to go extinct before their possible value to society is known. Every year, at a rate rather slower than that of plant extinctions but still very regularly, new and important uses for obscure and litle known plants are found. In the last 10 years the jojoba (*Simmondsia chinensis*) from the south west of the USA has become important for its oils, which are similar to those obtained from the threatened sperm whale, and it is now being planted experimentally as a crop.

Many endangered plants may be potential crops — at present the bulk of man's agriculture is based on less than 30 species. The Yeheb nut (*Cordeauxia edulis*), a bush from the Ethiopia-Somalia border is listed in the Red Data Book, but has great potential as a crop plant in the Sahel. Since its nuts have a relatively high protein content, it can grow in very arid regions, and it requires little in the way of cultivation.

The main threats come from agriculture, over-grazing, and felling for fuel and timber; industrialisation and pollution have also contributed. More recently, a number of plants are being threatened by horticulture. The increased prosperity of the developed countries has meant that many plant species are now exploited for non-essential uses and hobbies. ❜

*Earthscan Report*, Earthscan, 10 Percy Street, London W1.

**Our life is frittered away by detail. Simplify, simplify.** *H.D. Thoreau*

THE earliest recorded animal extinction was that of the European lion in about 80 AD. More than half of the extinctions that have occured since then have been in the 20th century. In the 350 years leading up to this century one animal species or sub-species per decade is believed to have disappeared. The International Union for the Conservation of Nature and Natural Resources (IUCN) estimates that currently the rate has stepped up to the disappearance of one animal species or sub-species every year.

In 1600 there were approximately 4,226 living species of mammals. At least 36 are now extinct, and 320 in extreme danger of extinction.

In 1600 there were about 8,684 living species of birds. Since then more than 100 have become extinct, and another 400 are in serious danger of extinction.

**continued over**

The food chain

The plants, animals, and micro-organisms that live in an area and make up the biological community are interconnected by an intricate set of relationships, which includes the physical environment in which these organisms exist. These independent biological and physical components make up the ecosystem.

Biological systems tend to concentrate poisons. For example, oysters have been found to accumulate up to 70,000 ppm (parts per million) of chlorinated hydrocarbon insecticides, many thousands of times the concentration found in their environment.

Food chains are "biological amplifiers". This tendency is especially marked in the case of chlorinated hydrocarbons because of their high solubility in fatty substances and their low solubility in water.

Losses of DDT and related compounds along a food chain are small compared to the amount that is transferred upwards through the chain, and the concentration increases dramatically at each level. Concentrations in birds at the end of the food chain are from tens to many hundreds of times as high as they are in animals further down the food chain. In predatory birds the concentration of DDT may be a million times as high as that in estuarine waters.

DDT is the oldest and most widely used chlorinated hydrocarbon insecticide. It is found everywhere-not only where it has been applied. DDT residues have been discovered in Antarctic penguins and seals.

Because DDT breaks down so slowly it lasts for decades in the soil. An estuary in Long Island, USA, was found to contain up to 32 pounds of DDT per acre in the upper layer of mud. This followed spraying of the area with DDT for 20 years as part of a programme to control mosquitoes.

The danger to fish-eating birds is extreme. Nesting failures among bald eagles have now reached the stage where the survival of the species is in jeopardy. DDT and other chlorinated hydrocarbon insecticides interfere with the birds' ability to metabolise calcium, which results in the laying of eggs with shells so thin that they are crushed by the weight of the nesting parents. Similar effects are caused by polychlorinated biphenyls, substances widely used in industry.

The publicity about the effects of DDT on falcon eggs helped spur severe restrictions on the use of chlorinated hydrocarbons in North America, and Europe. Yet persistent pesticides are increasingly and often profligately applied to Third World plantations.

---

**From Nature's chain, whatever link you strike, Tenth, or ten thousandth, breaks the chain alike.**
**Alexander Pope, *Essay on Man***

---

**World Wildlife Fund**
29 Greville Street
London EC1N8AX
Tel: 01-404 5691
An international organisation based in Switzerland with 27 national organisations. It was established in 1961 to raise money for the conservation of threatened wild animals, plants and places. WWFUK has five regional offices to coordinate the work of voluntary helpers and an educational branch — the Wildlife Youth Service.
Membership: £5.00 per year.

**continued over**

# ANIMALS and PLANTS

## The Clear Lake Lesson

IN an attempt to control the large number of gnats which populated Clear Lake, California, the area was sprayed in 1949 with DDD, a less toxic but equally persistent relative of DDT. The rate of application was about 0.2 ppm. The first application of what was then thought to be a relatively harmless pesticide eliminated about 99 per cent of the gnats, as did the next application in 1954. By the time the area was sprayed for the third and last time in 1957 the gnat plus about 150 other species of insect had developed some immunity to the pesticide. Within two weeks of each spraying no pesticide could be detected in the lake waters.

One of the first signs of ecological damage became apparent in 1950. Before 1950 Clear Lake had been a nesting ground for about 1,000 pairs of Western grebes. Not only did many grebes die soon after the 1954 and 1957 sprayings, but more died in subsequent years. The survivors were, moreover, unable to reproduce. From 1950 to 1961 no young were produced. In 1962 a single grebe hatched.

On investigation it was found that the concentration of DDD by the food chain had been the cause. The microscopic plankton of the lake contained about 250 times the concentration of DDD that had been applied to the water. The concentration in frogs was 2,000 times that of application; in sunfish, about 12,000; and in grebes, about 80,000 times.

It was therefore obvious why no DDD could be detected in the lake water soon after spraying. The insecticide's high solubility in fatty substances in biological systems had resulted in its being almost completely absorbed by the lake's living components.

## WHALES

IN 1933, 28,907 whales were caught, and they produced 2,606,201 barrels of whale oil. In 1966 57,891 whales were killed, but twice as many whales yielded only 1,546,904 barrels of oil, just about 60 % of the 1933 yield. This was one of the first indications of over-exploitation: as the larger kinds of whales were driven towards extinction the industry shifted towards harvesting not only young individuals of large species, but, with time, smaller and smaller species.

After the Second World War, the 17 countries interested in whaling established the International Whaling Commission (IWC). This commission was charged with regulating the annual harvest, setting limits to the catch and protecting the whale species from extinction. In theory, commissioners from the various nations were to be responsible for ensuring compliance with IWC decisions, but in fact their powers of inspection and enforcement were non-existent. Instead of setting quotas on the basis of "blue whale units". A blue whale unit (bwu) is one blue whale or the equivalent in terms of other species: two fin whales, two and a half humpback whales, or six sei whales. The nations engaged in whaling in the Antarctic were allowed a combined quota of 16,000 bwu. Since the blue whales were the largest, they were the most sought after. Up until about 1950, although blues continued to be taken, their numbers declined sharply, which meant that the next largest, the fins, were hunted more vigorously. Since the turn of the century at least 340,000 blue whales have been killed, 97 per cent of them in the Antarctic. In 1930 there were still an estimated 100,000 blue whales roaming the seas.

**The worst government is the most moral. One composed of cynics is often very tolerant and humane. But when fanatics are on top there is no limit to oppression.** *H.L. Mencken*

**continued over**

In the 1930/31 whaling season alone 29,410 were caught in the Antarctic. In 1960 the species was given partial protection. In 1967 members of the International Whaling Commission, including Japan and Russia, were banned from catching the blue whale. The Japanese promptly got around this by starting whaling companies in Chile and Peru — both non-member countries. Up to 1970 they were still killing blue whales under flags of convenience. The latest estimate indicates that there are between 17,500 and 19,000 blue whales remaining.

---

The International Whaling Commission has proved totally ineffective in checking the short-sighted greed of its members, with the result that some species are in serious danger of extinction.

---

A few species of whale are involved in a large volume of trade. Over the years the species concerned have changed, due to successive over-exploitation. Their long life, low reproductive rate and vulnerable life cycle make whales easy to overexploit and population recovery can take 10-100 years.

Species currently exploited on an industrial scale are the sperm whale, sei whale, Bryde's whale, minke whale and fin whale. The blue whale, once the mainstay of the industry, is now so rare that it is protected, as are the right whales, humpbacked whale and grey whale. Only the grey whale seems to have made any really significant recovery since it was protected.

The greatest destruction of populations of all species took place in the Antarctic. It was not until Antarctic whaling was no longer profitable that European countries began to abandon it.

In recent years the trade in whale products has declined, mainly because of reduced catches, in turn caused by reduced stocks.

Whales have (or have had) an extraordinary variety of commercial uses. Whale ambergris (from the intestine of the sperm whale) is used in scent and high quality soap. Baleen ("whalebone") is used in corsets, riding crops and umbrellas. Whale blood is used as fertiliser and as glue. Glands yield pharmaceuticals such as Vitamin A from the liver. The skin is used as leather for bicycle saddles, handbags and shoes. Sperm oil is used for dressing hides, hair oil, shaving cream, lubricating oil, printing ink and detergents. Tendons have been used for tennis racket strings and surgical stitches. Whale bone provides shoe horns and chess sets, and boiled whalebone gives gelatine for photography and jellies. Whale meat is used to feed pets, live-stock, zoo animals, mink on fur farms, screw-worm flies and humans. Whale oil can be used for dynamite, medicines, varnishes, linoleum, paint, margarine, lard, shortening, candles and crayons. Whale teeth provide carvings and piano keys. ❜

Earthscan

**Friends of the Earth, 9 Poland Street, London W1**

Aims: to reduce the impact on the environment of human activities, and to develop alternatives to ecologically harmful practices. Current campaigns on energy, wildlife, transport, resources, bicycles, land-use and agriculture.

# RESOURCES

THE RESOURCES THAT SUSTAIN US ARE OF THREE KINDS:

**The materials from which the crust of the Earth and the atmosphere are made;**
**The living things that inhabit the Earth, including ourselves;**
**The energy by which chemical reactions are made to occur, by which work is done and by which the Earth's biochemical cycles are powered.**

---

IF we seek to devise a way of life that can continue, in safety, into the remote future, then we would do well to base that way of life on the use of resources that will remain easily obtainable and plentiful for as long as the Earth continues to exist. Where we are dependent on resources that can be exhausted it would be wise to free ourselves from such dependence. It follows, then, than we would be wise to support and encourage those technologies that rely on materials that replicate themselves or that are recycled naturally within a reasonable space of time — say a century — or that are so common as to make exhaustion impossible.

### THE EARTH'S CRUST

THE most common elements in the Earth's crust, in descending order, with estimates for their amount in parts per million by weight are:

| Element | ppm occurrence |
|---|---|
| oxygen | 466,000 |
| silicon | 277,200 |
| aluminium | 81,300 |
| iron | 50,000 |
| calcium | 36,300 |
| sodium | 28,300 |
| potassium | 25,900 |

These elements may be regarded as very abundant and whatever other difficulties may be encountered in their exploitation, their exhaustion is unlikely.

The relative abundance of an element gives little indication of its usefulness, or of the ease with which it may be located.

Historically the first metals to be discovered and worked were those that occur in native (metallic) form, such as gold and copper. These metals were joined by tin, to make bronze, iron, silver, lead and, by Roman times, zinc which makes brass when alloyed with copper. Of the non-metallic elements, oxygen is present in rocks as a component in compounds with other elements, and it is obtained much more easily from the atmosphere or from water than from rock, while carbon and hydrogen are used in the construction of living tissue, and so conveniently isolated and concentrated.

## What does "depletion" mean?

EVEN in the case of rare substances it is misleading to imagine that they are likely ever to be exhausted, in the sense that they will be used up and none will remain. Unless they are transmuted to other isotopes (as is uranium used in fission reactors) or exported from the planet, elements cannot be lost and the stock of them must remain more or less constant. What are exhausted

**continued over**

| Element | ppm occurence | Element | ppm occurrence |
|---|---|---|---|
| magnesium | 20,900 | strontium | 375 |
| titanium | 4,400 | sulphur | 260 |
| hydrogen | 1,400 | carbon | 200 |
| phosphorus | 1,050 | zirconium | 165 |
| manganese | 950 | vanadium | 135 |
| fluorine | 625 | chlorine | 130 |
| barium | 425 | | |

# RESOURCES

**continued**

are those deposits in which substances are concentrated naturally in amounts that exceed by many times the average concentrations to be expected in rocks — that exceed the relative abundances predicted by the table of abundances above. It is now rare to find native copper, and gold which once lay at the surface must now be mined at great depths. As the most easily worked deposits are exhausted work must begin on more difficult deposits. These may be difficult because the desired substances are spread more thinly, or because the deposits are in remote or difficult terrain, or both. As the difficulty of working increases so does the cost, until a point is reached beyond which mining becomes uneconomic because the product cannot be marketed at an acceptable price.

---

The total global stock of a material is the "resource" and that part of it that can be extracted economically is the "reserve". The only reasonable way to calculate the size of a resource is to use the table of relative abundance in conjunction with the known weight of the Earth's crust. Reserves, however, are measured more carefully. Preliminary surveys may indicate geological conditions in which particular substances have been known to occur and so such surveys are used to produce estimates of "inferred reserves". No one can know whether in fact the substance is to be found in a particular place until holes have been drilled and samples of rock examined. If such detailed investigations show the presence of the desired substance an estimate is made of the quantity present and the resulting figure is the "indicated reserve". The amount cannot be known precisely until actual mining has begun and the estimated content of an actual mine is added to the "measured reserve".

The surveying, prospecting and mining are performed by mining companies and they are interested only in their own future. All of their operations are costly and they are not undertaken unnecessarily. If the measured reserves of a substance seem adequate to meet demand for, say, 30 years ahead, then there is no point in seeking further reserves. If it seems likely that within about 30 years new sources of supply will be needed, the search for them can be justified. In making such searches, however, companies will bear in mind any economic or political problems that could be encountered and so it may be that whole regions of the world will be omitted from their investigations, either because they are not permitted to enter those countries now or because they anticipate difficulties once mining has commenced. In recent years many Third World countries have been omitted from geological investigation because companies are unwilling to risk the nationalisation of their enterprises once capital has been invested in them. At the best, estimates of reserves are unlikely to extend more than 30 years into the future.

---

It is important to understand the way in which figures for reserves

are compiled, and why, if we are to use them correctly. At best they are likely to be very inaccurate if they are used to make predictions further into the future than they are intended to cover. Typically they will show the exhaustion of all the reserves after about 30 years! Nor can they be used to anticipate industrial behaviour even in the short or medium term. China might suddenly enter the world market in times of scarcity to sell materials no one knew she possessed.

The estimates can be used to anticipate changes in sources of supply if they show, for example, that certain mines will be worked out within a definite period.

**continued over**

# RESOURCES

**continued**

### Reaction to scarcity

WHEN a mineral deposit becomes uneconomic the mines working it close, but later technological developments may make it possible to open them again, or prices may rise sufficiently to make them economic even using their traditional methods.

---

Historically we have reacted to scarcity by searching wider or deeper. Today we can add to these the possibility of searching for much poorer grades of deposits. In the future it is possible that we may exploit extraterrestrial resources: this is the subject of study at present.

In some cases we have reached the practical limit by searching wider or deeper. We have located the best deposits in those parts of the world to which we have access and deep mines are often very deep indeed. The technology for working much more sparse deposits exists, however, and with some minerals it is already economic. The highest grade deposits of copper (55 ppm relative abundance), zinc (70 ppm), lead (13 ppm), tin (2 ppm), molybdenum (1.5 ppm) and mercury (0.08 ppm) are virtually exhausted. In the 1920s it was economic to mine copper ores with a copper content of about 1.5 per cent (which rose to about 2 per cent during the depression). Today ores containing about 0.75 per cent copper are being mined. It is possible today to mine iron from ores containing about 10 times more of the metal than do common rocks (i.e. 10 times more concentrated than the table of abundance figure), copper from ores containing 100 times more than common rocks, uranium from ores containing 1,000 times more than common rocks (which contain about 1.8 ppm), and mercury from rocks containing 100,000 times more than common rocks. It is not impossible that one day it will be possible to extract metals from the common rocks themselves, so that our "reserves" come to equal our "resources".

This suggests some future period of abundance, but if we are to enjoy it we must be prepared to pay a high price. If minerals are worked in this way the procedure — and it is the only procedure at present — is to remove the overlying soil to expose large masses of rock, cut out the rock, crush it on site, remove the partially refined metal, and leave behind the tailings. When rock is crushed air enters among the particles so that its volume increases, and the more or less powdered residue is too bulky to be returned to the holes from which it came. Clearly, the effect

HOW'S THAT FOR PRODUCTIVITY ?

on the landscape is both large and long-lasting, and the control of pollution from dust, noise, and from toxic materials that may be washed from the spoil, is difficult. In practice, attempts to strip mine in this way meet strenuous opposition from residents of the areas they will affect, and politically they are so unpopular that almost any alternative would be preferable.

The first alternative — although it is not really an alternative — is to increase the amount of material that is recycled and used again. The amount of recycling could be increased, but already metals are recycled on quite a large scale. The economics of recycling depend on the price for new material and usually recycled materials are inferior, because of the presence of contaminants. Recycling helps now, could help more, but it does not solve the problem.

Perhaps, then, we could use less by being more thrifty in our use of materials? Each year the world uses about 700 million tonnes of steel, 70,000 tonnes of antimony (0.2 ppm abundance), 16,000 tonnes of cadmium (0.2 ppm), 7 million tonnes of copper, 490 million of pig iron, 3.5 million

**continued over**

# RESOURCES

**continued**

tonnes of lead, 86,000 tonnes of molybdenum, 600,000 tonnes of nickel (75 ppm), 340,000 tonnes of titanium, 40,000 tonnes of tungsten (1.5 ppm), and 5.5 million tonnes of zinc. What is more, in most years the amount increases. We could economise in two ways: by making articles that last longer; or by freezing economic growth at its present, or some other, level.

Technically there is no reason why we cannot make goods that last for a long time. A cooker, refrigerator or heater should be capable of becoming an heirloom; a car should last a lifetime. Unfortunately our economic development has been such that our jobs and our prosperity depend on producing goods. If we are to continue to produce goods, then those goods must be used and replaced. Our evolution into the consuming society was implied from the start of the Industrial Revolution. If, now, we aim to make goods that are to be replaced much less frequently we must apply ourselves to the consequent problem of umemployment and economic stagnation. These problems may be surmountable — for one thing we might devote much more labour to the manufacture of each article and increase its price in proportion — but they must be faced and we must not pretend the transition from our present economic structure to the new one would be simple.

If we seek to freeze our level of economic activity we do not really touch the essential problem, which is the industrial development in countries that are just entering the phase from which we may be emerging. We cannot freeze their economic development and if we could the effect would be to freeze the present disparities between rich and poor countries.

It may be, though, that rather than attempting some heroic freezing of the world economy we might transfer it. If the industrial countries of Europe and North America, say, were to phase out certain industries in favour of

ALL
CHANGE
PLEASE

others (such as steelmaking, the motor industry and shipbuilding, in favour of, say, electronics, communications, education and entertainment) the effect might be to provide developing countries with the opportunities they need while keeping ourselves fully occupied. This would involve moving from an economy based on secondary (processing of raw materials) industries toward one based on tertiary (service) and quaternary (education, arts, etc.) industries. Indeed, this may be the historical process that we are experiencing at present.

**Resource substitution**

THE clearest possible evidence of the proximity of resource crises is the intensity of research into substitutes for those resources that may become scarce. In general, this research tends to employ as substitutes materials that are much more abundant generally than those they replace rather than simply to seek rich deposits of materials that may not be generally abundant. Copper, for example, is giving way to aluminium for some electrical uses and it is possible that electrical power may be conducted very efficiently through glass (made mainly from silicon). Copper and lead plumbing is giving way to plastic piping. Plastics are made from petrochemicals, which are also scarce, but the petrochemical feedstocks are also the subject for substitution studies. The ingredients of those feedstocks

**continued over**

# RESOURCES

**continued**

are organic compounds made primarily from carbon, hydrogen and oxygen, all of which are plentiful in living matter. Starchy agricultural crops, such as potatoes, sugar beet and sugar cane may make satisfactory substitutes. A substance made from carbon, glass and epoxy has been used (by the US Ford Motor Company) as a substitute for steel in car springs and reinforced plastics are being used as a substitute for mild steel in car bodies. Communications and the processing of information is becoming dependent on microprocessors made principally from silicon.

---

Such changes are sensible in terms of resource use and they should be welcomed, although with caution. The refining of aluminium, which never occurs naturally in the metallic form, is costly in energy and dirty, so that ways are still needed for improving the technology. Where agricultural crops are grown to supply feedstocks to the plastics and chemicals industries there is a risk that these enterprises will compete for land against food production, although on a modest scale they can be welcomed since they provide useful outlets for produce (such as sugar cane and beet, potatoes and possibly cassava) that are valuable to farmers but that are nutritionally undesirable or tend to be overproduced or both.

---

**Water**

FRESH water is an essential resource that is in short supply in many parts of the world, and it may be shortages of water that inhibit many industrial developments. Unlike mineral resources which are engaged in geological cycles lasting millions or hundreds of millions of years, the hydrological cycle is short and a particular molecule of water may be expected to complete it within about a year provided it does not become incorporated into the icecaps. Each year about 35 per cent of the total water on the surface of the Earth moves through the cycle. About 336 x $10^{12}$ tonnes evaporate from the sea and about 64 x $10^{12}$ tonnes from the surface of the land. This forms 400 x $10^{12}$ tonnes of cloud which deliver 100 x $10^{12}$ tonnes of precipitation over land and 300 x $10^{12}$ tonnes over the sea, leaving a balance of 36 x $10^{12}$ tonnes of water to flow as freshwater from the land to the sea (calculated as precipitation over land minus evaporation from the land). However, the proportion of the Earth's water that exists at any one time as fresh water in rivers and lakes is no more than 0.01 per cent and it is on this amount that we depend.

There is not much we can do to increase the size of this resource. If we extract groundwater there is a risk of lowering the water table. This may reduce the flow in rivers and it may permit seawater to enter the groundwater leading to problems of salination. We can distil seawater to obtain freshwater but unless we are prepared to pipe the freshwater to the dry interiors of continents desalination is likely to benefit only

those areas that lie in a narrow coastal belt. Even then, desalination produces as its by-product a brine whose salinity is much higher than that of the original seawater. To some extent the brine can be used as a source of mineral salts, but if it must be disposed of there can be difficulties. If it is discharged into the sea it will raise the salinity locally with adverse effects on marine organisms and the possibility that this super-salinated water will be drawn into the desalination plant where it will require additional processing. It can be stored in open lagoons above permeable structures so that additional freshwater is lost by

**continued over**

# RESOURCES

**continued**

evaporation and some water is partially purified by the structures through which it descends, but the only really satisfactory way to dispose of such strong brines is to dilute them with freshwater to about the salinity of the adjacent sea — which seems to defeat the purpose of desalination! This is not to dismiss desalination entirely. Locally and on a modest scale it can be very helpful, but it is unlikely to prove practicable on a large scale.

Existing water can be managed more efficiently, but in the long run ways must be found to economise in its use. At present the need for such economy is felt only in places or at times of drought, but eventually it may be felt everywhere since one way the problem might be shared is to export water from humid to dry areas. Manchester might become a net exporter of water, but if it were to do so Mancunians would need to use water more thriftily and to increase their storage capacity to ensure an exportable surplus.

**Living resources**

WE use plants and animals as sources of food, natural fibres, building materials and fuel. To a large extent their management is a matter for agricultural and forestry policies, but there are wider implications, relating especially to land use, energy policy and genetic conservation.

Our dependence on living resources is increasing rather than decreasing, and nowhere in the world can we afford to lose fertile land, even if such land is not being used at present. Land use policies must direct non-agricultural developments to the poorer regions in order to preserve as large an area of good agricultural land as possible. Where the loss of good land is inevitable, it should be minimised. Mineral extraction, for example, can take place only where the minerals are located. Mining companies are careful not to dispose of their wastes in such a way as to bury potentially workable mineral deposits beneath them. They should be required to avoid burying good farm land also. Railways are preferable to large roads for communications because they demand less land.

Fuelwood still accounts for a significant proportion of the world's primary energy sources, and much of this wood is reduced to charcoal, a process that is wasteful of the resource. The World Bank has estimated that at current rates of felling and replanting all the great forests of the world will disappear within about 50 years from now. Their loss would be undesirable for two reasons: timber has other uses than fuel and in years to come it may prove to be important as a source of organic chemicals from regions that are unsuited to other types of cropping; the large forests, and especially the tropical moist forests and montane forests include a significant part of the world's genetic resources that we can ill afford to lose.

---

Around the coasts of Europe there grows a small, rather insignificant plant, *Brassica oleracea*, the wild cabbage. It is the direct ancestor of all our brassica crops. Had it been made extinct by early man, we would not be able to eat cabbage, cauliflower, brussels sprouts, turnips or any of the other brassicas. We use only a tiny proportion of the species that inhabit the planet. There are many more we could use and with each species that we render extinct one possibility is closed to our descendants. These possibilities are very real. A few years ago the US National Research Council supported a study of 400 plant species that had never been domesticated. It was found that 36 of them, 9 per cent, were potentially useful and some of them were useful indeed. They included a bean with a very high protein content, another bean that yields an oil similar to sperm oil, a tree that grows on very salty soils and that could provide browse for sheep, and a shrub that produces more rubber than the rubber tree. We can protect the genetic resources of the planet only by the most vigorous conservation policies.

**continued over**

# RESOURCES

**continued**

## Fuel and energy

THE Earth and all the living things upon it derive their energy from solar radiation (light and heat) and from gravitational attraction and these forms of energy derive ultimately from the "big bang" which created our universe.

Traditionally we have utilised stored solar energy in the form of wood and, more recently, peat, coal and petroleum and more recently still we have learned to exploit the radioactive decay of some isotopes of uranium. In all of these we are accelerating a natural process. The organic fuels are obtained from materials that will oxidise as part of their decomposition and by burning them we accelerate the oxidation with the release of heat and light. The fossil fuels are obtained from materials whose oxidation has been arrested. By burning them we complete it. Uranium decays naturally. By concentrating it and amassing it in reactors we provide conditions under which the natural decay is accelerated.

The use of wood for fuel can be sustained provided the rate of consumption does not exceed the rate of new growth for any length of time, but in the modern world this reduces the role of wood to that of a relatively unimportant source of energy and for reasons of conservation it may be desirable to phase it out almost completely.

FOREST
WALK

The usefulness of other fuels depends on their availability.

As with all mineral resources, figures for reserves must be treated with caution for new discoveries continue to occur. Since the first fears of oil depletion were popularised there have been new fields discovered in Alaska, the North Sea, under southern England and in Mexico there are reserves that may be comparable to those of Saudi Arabia. During the same period very large coalfields have been discovered in England.

The limit to our rate of consumption of fossil fuels is determined not by the size of the reserves but by the capacity of pumps, pipelines, refineries and ships to extract, process and transport it. In the long term, though, it is inevitable that these limits will be sufficient to increase prices to levels at which it becomes uneconomic to burn petroleum. Coal will last much longer, but its principal disadvantage is the cost and relative inefficiency of the processes that convert it to the liquid fuels we need to power mobile engines. Peat forms continuously, but on a scale that is too small to make peat a fuel with more than local value.

Clearly we need alternatives to fossil fuels, although this need may be less urgent than it seems. The official alternative is to move to uranium. Uranium is also a limited resource and one that is very expensive to process and use. Quite apart from the inherent dangers of the nuclear fuel cycle there are good grounds to doubt whether nuclear power can ever be profitable on a large scale. At best it can produce only base-load electricity, with which most countries are adequately supplied. What is needed is rather different. We need peak electricity to supply short-term, local high demands, liquid fuels, and chemical feedstocks. Nuclear power cannot supply these. The cost of nuclear power is so high that should we invest in it heavily there is a real risk that the drain on capital resources will deprive other industries of needed investment and alternative energy sources of funds for research and development.

The energy problem is unique in that far from lacking solutions it has so many solutions that we are forced to choose.

**continued over**

# RESOURCES

**continued**

Immediately, much greater emphasis should be placed on conservation of energy. This would require the insulation of buildings to much higher standards and certain developments in recent years may help in this. A family of clathrates have been developed that freeze and melt at about ambient temperatures with the absorption and release of as much latent heat as water. A clathrate is a molecule that has a lattice structure and that encloses another, quite different molecule. In this case the enclosed molecule is water. Clathrate insulators might consist of sealed containers placed inside cavity walls. As temperature rises the insulation absorbs heat, as the temperature falls heat is released. Research is now very advanced into alternative sources of energy. Ethanol is being produced on a commercial scale in Brazil for mixing with petrol. The production of hydrogen from water is becoming cheaper and the problems of using hydrogen as a liquid fuel are being solved. A hydrogen-powered aircraft is being designed. The solar cell is almost competitive for small-scale electricity generation and its price is falling. All of these are additional to the more obvious sources such as wind, water, wave motion, tidal motion and solar heat.

These new sources of energy deserve support and encouragement. They are based on resources that are either inexhaustible (solar heat, solar light, water gravity) or that renew themselves. At the same time they are benign in their effects. Solar cells are made from silicon with the addition of other common materials such as boron and phosphorus. In use they would be mounted on roofs where they would be inconspicuous. Ethanol has a high octane rating so that when used as a fuel in existing engines it is not necessary to add tetraethyl lead, so that an ethanol powered engine causes less pollution. When hydrogen is burned as a fuel the by-product is water vapour.

WE should beware of any figures that purport to be firm estimates for the size of any resource and we should remember to distinguish between the concepts of resources and reserves.

We should recognise that non-renewable resources are being used rapidly. This does not mean that they will disappear from the Earth, but it does mean they are likely to become more and more expensive.
We should encourage any attempt to switch from a scarce resource to one that is more plentiful. Where this creates new environmental problems we should try to find solutions to these problems — perhaps delaying the switch until they have been found — rather than simply opposing them, and we should compare the problems with those of the alternatives. Aluminium processing is environmentally unpleasant, for example, but arguably the mining of low grade copper ores and their refining is worse.

# FOOD

**He gave it for his opinion, that whoever could make two ears of corn or two blades of grass to grow upon a spot of ground where only one grew before, would deserve better of mankind and do more essential service to his country than the whole race of politicians put together.**
**The King of Brobdingnag in** *Gulliver's Travels*

## What is efficiency in food production?

In places like North America and Australia where there is abundant land, energy and machinery but expensive manpower, efficiency has been measured by production per man-hour.

In places like England, Holland, Denmark and New Zealand where energy and machinery are cheap but land is in short supply and manpower is expensive, efficiency has been measured not only by production per man-hour but also per acre..

By these standards the countries so far named have the most efficient agriculture in the world.

But there is another way of looking at it. This is to measure efficiency in terms of the input of energy compared to the output of food. By that standard North America, England, Holland and Denmark are extremely inefficient.

'The smallest farms in Brazil and Argentina are eight times as productive as the biggest ones, and in Colombia they produce *fourteen* times as much per acre. Of course, big farms usually give higher productivity *per man,* but poor countries need to get as much food as possible out of every *acre.* Labour is no problem, for there is massive underemployment, most of it in the countryside and much of it among the small farmers and their families.'

Geoffrey Lean,
*Rich World, Poor World,*
Allen and Unwin 1978

WHEN agriculture began in the Middle East about 10,000 years ago the early farmers used only the energy of the sun, their own muscle-power and machinery no more complicated than a flint-edged sickle. By these means each worker could produce enough to feed himself and another fifteen to twenty people. The energy gain was therefore between 1:15 and 1:20.

Modern food production in North America and Western Europe uses so much oil and other fossil fuels at every stage of the process — from making the machinery and artificial fertilisers to packaging, distribution and merchandising — that it shows a clear energy loss. The input is actually bigger than the output. The ancient Sumerian farmed about 20 times as efficiently as the Agribusinessman.

**"Nay, sir, whatever may be the quantity that a man eats it is plain that if he is too fat he has eaten more than he should have done."**
*Samuel Johnson*

❛ So distorted has the ratio become that the potato in the shopping basket is wholly made of oil (in the energy sense) and the modern battery-produced egg probably uses six times as much energy in its production as it offers in food value . . . When a field of wheat is harvested or an egg collected from a battery hen, a huge energy debt has already been incurred. Tractors and combines have been built, fertilizers and pesticides have been synthesized, new strains of crops and farm animals raised, heavy expenditure on heated buildings, crop dryers, processing machinery, transport and egg packaging has already been committed. Although the yields per acre, or per hen for example, have increased greatly in the last hundred years and dramatically in the last twenty, these increases have been bought by the expenditure of enormous quantities of energy. ❜

Barry Juniper.
*The Countryman*
Autumn 1977

# FOOD

**continued**

In terms of turning money into more money the giants of Agribusiness are highly efficient.

In terms of converting energy into food they are among the least efficient farmers in the world.

"The most successful farmers in the Western world, in strict monetary terms, are those who have mechanized, i.e. those who import the most energy. This apparent paradox comes about because the intelligent farmer can do far more work with £39-worth of diesel oil than can a single farm labourer in a week (this being the official minimum weekly wage, 1977). But we are using that oil a million times faster than it was made. What farmer would squander his capital assets so wantonly and what price can be put upon such an evanescent resource?" Barry Juniper

For really efficient food production you want to look at the back gardens and allotments. Compared with agribusiness they produce little per man-hour. On the other hand, they produce a colossal amount per acre and per unit of energy. They involve virtually no machinery or transport, no packaging or merchandising. If there's a compost heap, there's virtually no waste either.

ONE does not need to be a vegetarian to recognise that the diet of the Western world relies far too much on meat-eating. Meat production on a vast scale leads to cruel farming practices. In addition to moral objections, there are sound and pressing practical ones.

Meat is an extremely inefficient way of producing protein. Grass turns 1 per cent of the sun's energy into food. The cow that eats that grass turns only 10 per cent of that 1 per cent into available food: the other 90 per cent is wasted.

**Vegetables give a far more efficient return.**

On one acre of land a man can produce enough protein to supply himself

for 2,224 days if he grows soybeans
for 887 days if he grows wheat
for 354 days if he grows corn
for 77 days if he produces beef.

To put it another way, it needs nearly five acres to provide protein from beef for one man for one year. The same area producing soybeans would provide enough protein for more than 30 people.

---

**"Tell me what you eat and I will tell you who you are."**
**Anthelme Brillat-Savarin 1825**

---

**The efficiency of protein conversion**

The soya bean turns sunlight into protein without any help. This is a 100 per cent gain in protein. To get protein from cattle you have to feed in large quantities of pasture, fossil fuel energy, fertilisers and so on. The table shows the relative efficiency of protein conversion of various forms of food.

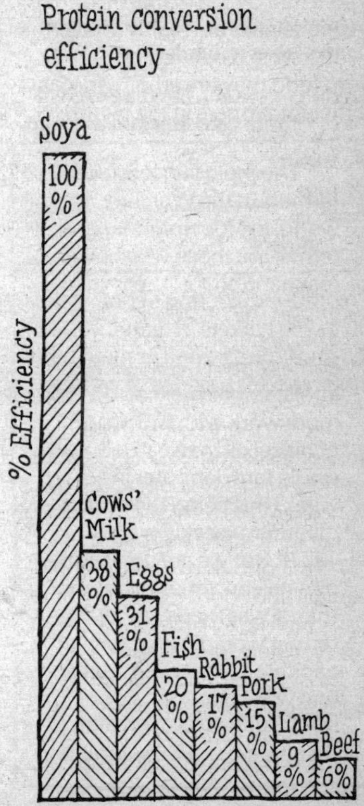
Protein conversion efficiency
% Efficiency
Soya
100 %
Cows' Milk
38 %
Eggs
31 %
Fish
20 %
Rabbit
17 %
Pork
15 %
Lamb
9 %
Beef
6%

# FOOD

continued

**In Britain about half our protein comes from animals.**

**In the course of a life time each of us in Britain consumes on average**

**56 sheep**
**36 pigs**
**8 cattle**
**550 poultry**

To produce a pound of grain takes 60 to 250 gallons of water.
To produce a pound of meat takes 2,500 to 6,000 gallons of water.

The Northern hemisphere consumes about 90 per cent of the world's oil production, about 80 per cent of the fertilisers and about 75 per cent of the fish catch.

About one-third of the world's cereals and about a quarter of the fish catch is fed to the livestock of the Northern hemisphere.

Most soil is deficient in nutrients essential for efficient plant production. The price of higher food production is therefore higher fertiliser consumption.

There is a world shortage of inorganic (or artificial) fertiliser.

**"The United States uses more fertilizer on its golf courses, back gardens and cemeteries than is used for the total agricultural programme of India." McGraw** *Population Today* **Kay & Ward**

Britain imports about half of its food, at a cost of some £6,500 million. But it's worse than that, because much of our home-produced food depends on imported feeding stuffs, fertilisers and oil

A century ago in Britain and the United States we ate less than four pounds of **sugar** each per year. Now the average British or North American man, woman or child gets through about 2 pounds a week. Sugar supplies energy, but has no vitamins or minerals.

It also destroys teeth.

THE glorious **cabbage.** It produces 4 times the energy and 22 times the protein as pigs on the same acreage of land. It is 16 times more efficient than sheep, 8 times better than chickens, 10 times better than beef cattle at protein.

The mighty **spud** produces less protein than the cabbage but is three times as energy-giving. The potato is cheap, easy to grow, the food of poor people. It can grow anywhere except the low-lying tropics. It grows inside the Arctic Circle in southern Greenland, in central Alaska, in northern Scandinavia. Unlike wheat or rice it requires no complicated processing to render it edible. Just dig it from the ground and put it in a pot.

You can live on nothing but spuds. A century and a half ago, Irish peasants used to eat an average of eight pounds a day each, and they did a hard day's work on it. As a diet in itself, the spud is less monotonous than you might think. Wheat could be much worse. Without adding a thing to a potato you can bake it, roast it, boil it, steam it or fry it, and you have five dishes that are quite different from one another — six if you allow that boiled potatoes acquire new qualities when you mash them. Let the cook add more ingredients to the basic raw material and the list of dishes is virtually endless.

Potatoes are highly nutritious. More than three-quarters of their weight is water, but even allowing for that, a raw potato contains just over 2 per cent protein. This is much less than any of the grains, but because the crop yield is so much greater, you can grow about 420kg of protein per hectare with potatoes, compared with 350kg with wheat and only 200kg with peas. The protein in potatoes is well balanced, too, a nice blend of the amino acids the human body

**continued over**

# FOOD

**continued**

needs, so that little is wasted. If you fry the potatoes you will get rid of much of the water. This means you will eat a more concentrated food, with a protein count of about 3.8 per cent. Potatoes also supply the average British household with about one-third of its total supply of vitamin C. This is more than is supplied by any other food and more than we obtain from all the fruit and nuts we eat added together. ❜
Michael Allaby, *Vole* 6, 1978

---

**Rice** is the staple food of an estimated 2 billion people. Mainland China grows about 35 per cent of the world's rice, India and Pakistan 27 per cent, Japan 7 per cent and Indonesia 6 per cent. Other countries, largely in South East Asia and Latin America, grow about 25 per cent.

**Wheat:** the US produces some 15 per cent of the world's wheat, the Soviet Union 24 per cent, Canada, France and India 5 per cent each, Italy 4 per cent, Turkey 3 per cent, Argentina and Australia 2.5 per cent, and all the rest of the world about 34 per cent.

**Corn or maize:** over 50 per cent is grown in the US; Russia grows slightly over 5 per cent and Brazil a little under 5 per cent. A little over 13 per cent is accounted for by Yugoslavia, Mexico, Argentina, Rumania, and South Africa combined. The other countries grow 25 per cent jointly.

---

## FURTHER READING


Lawrence D. Hills *Organic Gardening* Penguin
N. W. Price *Food Resources* Penguin
Kenneth Mellanby *Can Britain Feed Itself?*
Chris Wardle *Changing Food Habits in the UK.*
Pete Riley: *Economic Growth: the allotments campaign guide* Friends of the Earth
Susan George: *How the Other Half Dies: the real reasons for world hunger.* Penguin
*The Vegetable Garden Displayed.* Royal Horticultural Society


---

**Make a compost heap**

Buy or make a wooden box to hold the compost. A double box is even better. Then you can have one on the go, and the other for use. Even better, you can turn the contents of one into the other. This turns the compost upside down and speeds the process.

Start with rough waste (hedge clippings, for example) which won't block the air vents. After that — garden rubbish, dead plants, leaves, cuttings, paper (shredded, and not too much), lawn mowings, animal manure, kitchen scraps. Anything organic: if it lived once it can live again. Build up layer by layer. About every ten inches sprinkle on lime, organic manure (pigeon droppings are rich in nitrogen, and may be available from local pigeon fanciers: ask the local pigeon club; Human urine is also rich in nitrogen), wood ash, sulphate of ammonia, or one of the proprietary brand activators.

Don't tread it down. When it's full give it a waterproof cover. An old carpet is ideal. If it gets very dry, give a sprinkling of water, but don't drench it. When it has gone hot and then cold it should be ready.

An excellent leaflet on compost-making is available free if you send a stamped addressed envelope to the **Henry Doubleday Research Association,** Convent Lane, Bocking, Braintree, Essex.

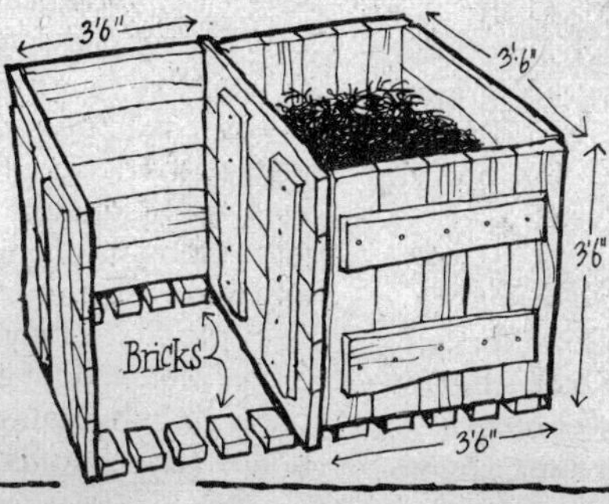

3'6"
3'6"
3'6"
Bricks
3'6"

# FOOD

**continued**

### Technological fish

❛The modern fisherman hunts with all the paraphernalia of modern technology. Huge trawlers track down their quarry with electronic devices, hauling in ton after ton of fish to be quick-frozen by the accompanying factory ship. Back in port skinned fillets are moulded and frozen into solid blocks of fish, then sawn into steaks to boil in the bag or cut into strips. The strips progress through a curtain of batter, a shower of brightly coloured crumbs, a half-minute dip in a tank of hot fat and finally another quick freeze down to —40 F in a blast of cold air. Ten fish fingers weigh about 220 grams, of which about 150 grams are water, 35 protein and 20 fat. Each fish finger, therefore provides nearly 40 Calories as well as an unidentifiable texture. To label them fish is an affront to one of nature's most wholesome foods.❜

George Seddon and Jackie Burrow: *The Wholefood Book* Mitchell Beazley

---

**Some ha meat, and canna eat,**
**And some would eat that want it;**
**But we have meat and we can eat,**
**And sae the Lord be thankit.**

*The Selkirk Grace, attributed to Robert Burns*

---

*The Two Faces of Malnutrition* by Erik Eckholm and Frank Record is one of the invaluable Worldwatch Papers, published by the Worldwatch Institute, 1776 Massachusetts Ave NW, Washington DC 20036.

One face of malnutrition is that of famine and undernutrition. *"undernutrition probably contributes to more than half of all child deaths in Latin America; a comparable calculation in South Asia or Central Africa would likely yield an even higher figure."*

The other face of malnutrition is the over-feeding and unhealthy diets of the rich countries. "Those with an affluent diet consume large amounts of animal proteins and fats in the forms of meats and dairy products; they substitute highly refined flour and sugar for bulky carbohydrates like whole grains, tubers, fruits, and

Useful Addresses

**Oxfam**
274 Banbury Road, Oxford Tel: Oxford 56777

**War on Want**
467 Caledonian Road, London N7 Tel: 01-609 0211

**The Vegetarian Society**
The Vegetarian Centre and Bookshop, 53 Marloes Road, Kensington, London W8 6LA Tel: 01-937 7739

**Compassion in World Farming** Lyndum House, Petersfield, Hampshire Tel: Petersfield 4208

**Henry Doubleday Research Association** 20 Convent Lane, Bocking, Braintree, Essex Tel: Braintree 24083

**Comet**
11 Harmer Lane, Digswell, Welwyn, Herts

**Ministry of Agriculture, Fisheries and Food**
3 Whitehall Place, London SW1A 2HH Tel: 01-839 7711

**The Soil Association**
Walnut Tree Manor, Haughley, Stowmarket, Suffolk

**Working Weekends on Organic Farms**
19 Bradford Road, Lewes, East Sussex

---

vegetables; and, increasingly, they choose commercially manufactured foods over fresh, unprocessed products."

Not only is the affluent diet wasteful of energy.
It is also unhealthy, and is thought to be largely responsible for the high incidence of heart disease, obesity, hypertension and cancer in the affluent countries.

*"Governments, through their agricultural, economic, and educational policies, have at best usually ignored the problems of overnutrition. At worst they have actively promoted unhealthy consumption trends . . . Thus, faced with mountains of surplus butter, the European Economic*

**continued over**

# FOOD

**continued**

*Community Commission recently proposed taxing edible oils to make margarine as expensive as butter — in effect, to encourage higher consumption of saturated fats. In Great Britain a recent Government White Paper on national food-production policy ignored health considerations while calling for increased output of milk, beef, and sugar beets. The Congressionally-mandated involvement of the US Department of Agriculture in promoting higher consumption of eggs by Americans provides another such example."*

*"Among the industrial countries, Sweden and Norway stand alone in their recent decisions to try to integrate modern dietary health concerns into national economic and agricultural planning. Particularly through a vigorous public education programme, the Swedish Government has worked to reduce the amount of calories, fats, sugars, and alcohol Swedes consume and to increase the amount of exercise Swedes get. The Norwegian Government hopes to establish a broad array of subsidies, grants, price policies, and other incentives that will promote a stabilization of meat consumption (which has been rising over the last decade); an increase in fish consumption; a reduction in feed-grain imports; a preference for low-fat over whole milk; a reversal of the decline in consumption of grains, potatoes, and vegetables . . . If the plan is implemented, Norway may not only better the health of its populace, but also reduce its agricultural trade deficit and reduce its claims on world food resources."*

**Worldwatch Papers** are available in the UK from Conservation Books, 228 London Road, Earley, Reading, Berkshire.

**You can travel fifty thousand miles in America without once tasting a piece of good bread.**
*Henry Miller*

## ALLOTMENTS

Friends of the Earth reckon that since 1970 the waiting list for allotments has increased by 1,600 per cent. In the same period the acreage and number of allotments has declined.

Shortage of land is no excuse. There are vast amounts of derelict land that could be cultivated, even in inner cities

*"As a nation we are now spending £1.45 billion more on fruit and vegetables than seven years ago. One allotment can save £130 a year on the family bill for fresh produce . . . Since Friends of the Earth first launched an allotments campaign in 1974, the amount of waste land has increased and the number of allotments fallen. In ten years the amount of waste land quadrupled in parts of London."* Friends of the Earth. 1979

**The capacity for progressive action on the part of any local authority is in inverse ratio to the need for it.**
*Old Kentish Aphorism*

During the Second World War the Dig for Victory campaign raised the number of allotments to 1,400,000 which produced about one-tenth of the food then grown in the country.

Friends of the Earth's booklet *Economic Growth: the allotments campaign guide,* by Peter Riley (£1.40 + p.&p. from FOE, 9 Poland Street, London W1) argues that the government should

1. Survey all waste land in the UK
2. Create a comprehensive and public register of land ownership.
3. Give local authorities the power to licence temporary allotments on any land idle for five years.
4. Instruct local authorities to clear their allotment waiting lists within two years by reclaiming waste land.
5. Instruct the Manpower Services Commission to initiate and support land reclamation projects.

# LAND USE

"THIS crowded little island" . . . "the disappearing countryside". What is the truth behind such much-used phrases?

ROLLING
ACRES

Middle-aged people remember with hazy nostalgia the way the country was in their childhood, when there were fewer machines, and more butterflies, and more wild flowers, and it was always a summer's day.

The reality was rather different. Agriculture in the 1930s was uniquely depressed. The city-dweller's romantic view of the joys of the countryside ignored widespread poverty among the families of farm workers. What appeared to be picturesque actually consisted of dilapidated buildings, and ill-cared for land. A field of buttercups or cornflowers may delight the city-dweller's eye: to a farmer it's just unproductive land.

For centuries people have bemoaned the passing of the countryside. What they are probably regretting in fact is the passing of their own youth. As Raymond Williams has pointed out, the process can be traced back a long way. D. H. Lawrence regretted the passing of the English countryside that Hardy knew. Hardy regretted the passing of George Eliot's England. Ruskin thought the countryside was being destroyed. So did Wordsworth. In the eighteenth century Goldsmith was movingly lamenting the fate of "Sweet Auburn, loveliest village of the plain." And so on.

The same sort of nostalgia has also been felt by city-dwellers for the environment of *their* childhood when neighbourhoods were neighbourhoods, and people talked to one another. Disentangling actual deterioration from middle-aged pessimism is not easy.

THE total land surface of the United Kingdom comprises 24.1 million hectares (59.6 million acres).

79 per cent are devoted to agriculture. 8 per cent are devoted to urban development. 7 per cent are devoted to forestry and woodland. The rest to miscellaneous purposes. Much of the agricultural land, especially in Scotland and Wales, is given to rough grazing. If this is not included then only half of the UK land surface is used for agriculture.

| Agricultural Land uses 1976 | Hectares 000s | Acres 000s | % |
|---|---|---|---|
| Crops and fallow | 4802 | 11866 | 19.9 |
| Temporary grass | 2314 | 5717 | 9.6 |
| Permanent | 4950 | 12232 | 20.5 |
| Rough grazing | 6768 | 16725 | 28.1 |
| Other land | 312 | 770 | 3 |
| *Total Agriculture* | 19146 | 47310 | 79.4 |

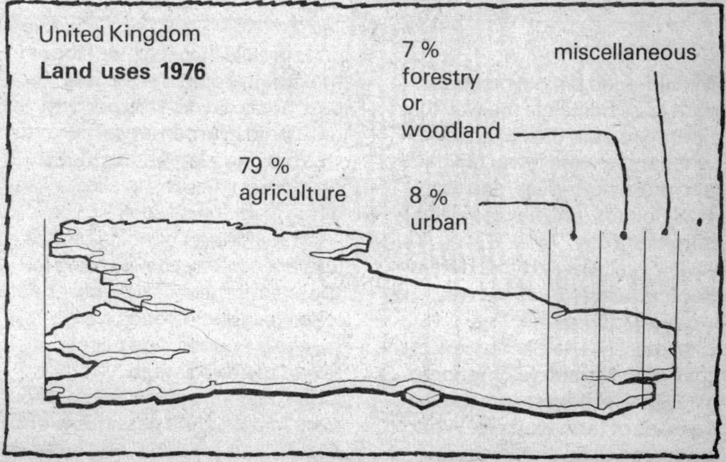
United Kingdom
Land uses 1976
79 % agriculture
7 % forestry or woodland
8 % urban
miscellaneous

# GROWTH OF THE CITIES

IN the past 100 years the population of the UK has more than doubled. In the middle of the eighteenth century it was 6 million. In the middle of the nineteenth century it was over 20 million. It is now 56 million and appears to have settled down at around that figure.

In the second half of the nineteenth century the 20 or 30 major towns and cities mostly doubled in population, though not always in area. They were tightly packed.

**We are an overwhelmingly urban population. Even so, on this crowded island, there is an acre of land for each of us, and about half of it is good agricultural land.**

In 1830 75 per cent of the population lived in rural areas. By 1920 75 per cent lived in urban areas.

# LOSSES OF FARMLAND

THOUGH we do not now lose the same quantities of agricultural land that were gobbled up by suburban sprawl in the 1930s, there is still considerable urban expansion.

The pressure on land in Britain this century has come not so much from population growth as from increasing land-take per person (up to 63 per cent). Land taken for urban development has increased 245 per cent.

Just in the decade 1961 to 1971 urban development on virgin land covered an area the size of Greater London.

FOR SALE
BY AUCTION
FERTILE BUILDING
LAND

Annual average net losses of agricultural land to urban use

('000 hectares)

| Period | England & Wales | Period | England & Wales | Period | England & Wales |
|---|---|---|---|---|---|
| **1922-26** | 9.1 | **1939-45** | 5.3 | **1960-65** | 15.3 |
| **1926-31** | 21.1 | **1945-50** | 17.5 | **1965-70** | 16.8 |
| **1931-36** | 25.1 | **1950-55** | 15.5 | **1970-74** | 15.4 |
| **1936-39** | 25.1 | **1955-60** | 14.0 | | |

# LOSS OF LAND TO ROADS

**Loss of land to roads**

MOTORWAYS take up, on average, 25 acres of land per mile. When junctions are included the land-take rises to 42 acres per mile. Completing the planned 2,000 mile motorway network plan would consume 22,000 acres of chiefly agricultural land.

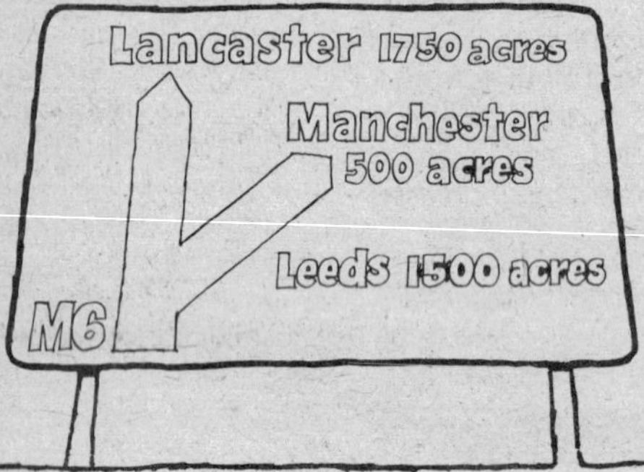
Lancaster 1750 acres
Manchester
500 acres
Leeds 1500 acres
M6

# THE WASTED LAND

MISUSE of land is rivalled only by non-use of land. Think of all the odd bits of our towns and cities that are hidden behind corrugated iron fencing (at £7 a yard). Think of the disused railway sidings, the odd bits around housing estates (known as SLOAP, Space Left Over After Planning).

*Urban Wasteland* (a report prepared by Timothy Cantell for the Civic Trust, 17 Carlton House Terrace, London SW1) estimates that this kind of derelict and unused land totals some 250,000 acres — an area the size of Birmingham, Derby, Glasgow, Hull, Liverpool, Manchester, Nottingham, Portsmouth and Southampton combined.

According to the report 34 per cent of this waste land is owned by local authorities, 6 per cent by British Rail, 24 per cent by private firms and developers.

Between 1962 and 1972 wasteland in the Thames Estuary has doubled.

Dr Alice Coleman, of King's College, London, is in charge of the Land Use Survey there. She maintains that modern planning has done little to reduce urban sprawl and its intrusion into the countryside. She argues that the waste of agricultural land not only reduces our ability to feed ourselves but also severely damages the landscape.

She has found that between 1933 and 1963 half a million hectares of improved farmland were lost. At that rate of loss all agricultural land would be gone in 600 years. In fact the loss was not spread evenly over the 30 year period, since the area of agricultural land increased during the war. The real loss occurred mostly in the 1950s and 1960s, and at that rate all agricultural land would be gone in only three centuries.

A disproportionate amount of the lost farming land has gone to road use, or waste. Urbanisation has taken place largely on land that is classified in terms of agricultural terms as first class or good.

**At the current rate at which agricultural land is disappearing, we are losing an area half the size of the Isle of Wight every year. Much of the land that is lost is of the highest agricultural quality.**

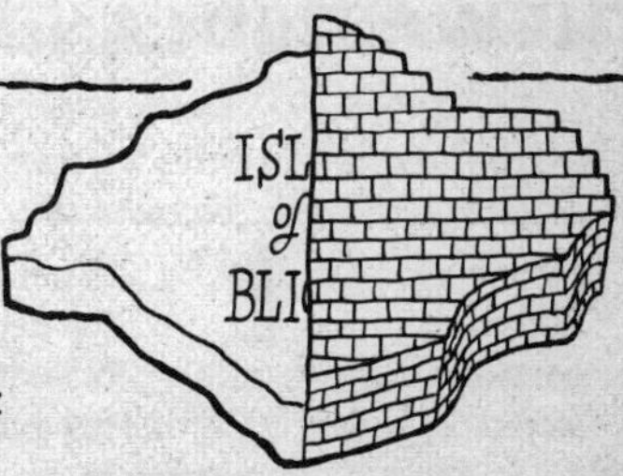
ISL
of
BLI

# THE BEST GOES FIRST

THE quality of agricultural land is assessed as to whether it is Grade 1 or 2, which is the land with the greatest productivity for a wide range of crops; Grade 3, valuable agricultural land for cereals, roots and grass to moderately good for oats, barley, roots and grass; Grade 4 and Grade 5, which are described as of "restricted potential".

Only 12 per cent of our farmland is in the top two grades, and less than half is in the top three.

Ministry of Agriculture, Fisheries and Food (MAFF) reckoned in 1976 that our farmland is distributed by grade in the following proportions

| | |
|---|---|
| Grade 1 | 1.8 per cent |
| Grade 2 | 9.8 per cent |
| Grade 3 | 36.4 per cent |
| Grade 4 | 18.2 per cent |
| Grade 5 | 33.7 per cent |

It is the best land that is most misused. It is there that most of the waste land occurs (First class land 4.3 per cent, Poor land 3.8 per cent). Again, the urbanisation takes places disproportionately on the best land.

**Newly urbanised areas in England and Wales (1933-1967) as defined by MAFF**

| Grade | National area | Newly urbanised |
|---|---|---|
| First class | 7.5 per cent | 9.6 per cent |
| Good | 37.6 per cent | 43.5 per cent |
| Medium | 36.5 per cent | 37.4 per cent |
| Poor | 18.4 per cent | 9.5 per cent |

FIVE ACRE
FARM

# THE PEOPLE  WHERE WILL THEY GO?

Vitality
Overcrowding
Public Transport
Stressful Lifestyle
Divorce from Nature
Parks and Public Spaces
Few Houses with Gardens
Social and Business Opportunities
Recreation and Cultural Activities
Traffic Noise and Pollution
Prospects of Employment
Impressive Architecture
Long Journey to Work
Anonymity and Privacy
Expensive Housing
High Land Costs
Vandalism

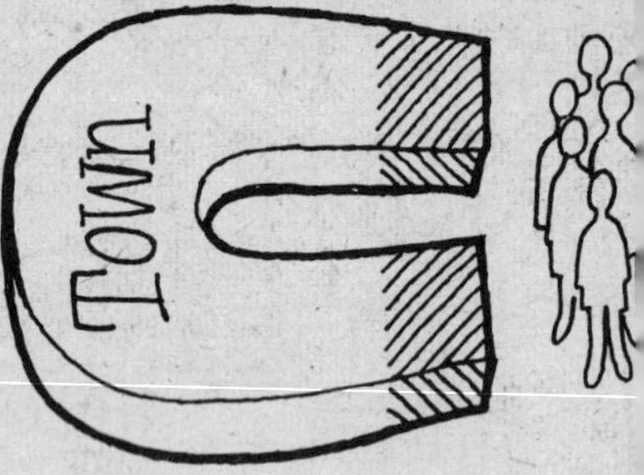
Town

---

This diagram is based on the Town and Country Planning Association's updated version of Ebenezer Howard's Three Magnets — the most famous planning diagram in the world. In 1898 Ebenezer Howard, a 48-year-old shorthand writer and Parliamentary reporter, expounded his idea of garden cities in a book called *Tomorrow: a peaceful path to real reform* (later re-published as *Garden Cities of Tomorrow).*

Words were translated into deeds with startling speed, as people caught onto the idea of places that could combine the advantages of town and country without the drawbacks (at least this was the ideal). Letchworth got going in 1903, Welwyn Garden City in 1919. The idea caught the imagination of the country and of the world, and has been imitated widely throughout Europe, in Czechoslovakia, Poland, Russia and North America.

Country

Personal Identity
Contact with Nature
Need for Land Reform
Lack of Social Services
Not Attractive to Young People
Lower Incomes and Fewer Jobs
Pressure from Tourists
Long Journey to Work
No Public Transport
Houses with Gardens

| | |
|---|---|
| Few Shops | Peace and Quiet |
| Fresh Food | Human Scale |
| Relaxed Lifestyle | Fresh Air |

Town-
Country

Sense of Community
Contact with Nature
All Forms of Transport
Adequate Social Services
Land Profits to Community
A Sense of Place and a Fuller Life
Recreation and Cultural Activities
Opportunities for Employment
Parks and Public Spaces
A Planned Environment
Convenient Shopping

| | |
|---|---|
| No Noise | Houses with Gardens |
| Work near Homes | Popular Involment |
| Relaxed Lifestyle | No Pollution |

Ebenezer Howard's ideas are now propagated by the **Town and Country Planning Association,** an educational body supported by voluntary subscription. TCPA, 17 Carlton House Terrace, London SW1. Tel. 01-930 8903/4/5.

# HOW TO SAVE DERELICT LAND

1 Identify the eyesore idle plot

2 With a local amenity society, or Friends of the Earth group, or simply off your own bat, establish the ownership of the land. Ask neighbouring owners, or the District Council's planning or rating section.

3 Find out from the District Council planning department if any planning permission has been granted

4 If it has, and it's a good scheme, campaign to have it implemented. If it hasn't, then put forward your own ideas.

5 Write to the local paper denouncing the eyesore. Or get the news editor to run a story on it. If you have a scheme, get sketches of the site BEFORE (as it is) and AFTER (when your scheme — play park, urban farm, or whatever has been implemented). Or get the local paper to sponsor a competition to come up with the best proposal for the site.

6 Apply to the District Council for planning permission for your scheme. You are risking no liability by doing so. If it is granted, you don't have to carry it out, but it would serve as a useful pressure on the owner.

---

"FOR a more general campaign on waste land, the springboard will be a survey — of the whole of a village or small town or a part of a larger town. This would identify every dormant site on a map with a schedule giving such basic facts as ownership, size, condition, planning permission and suggested uses. Students or school-children might be invited to help with the fieldwork. The District Council planners might help with maps (and an unsympathetic response to a letter to the Chief Planning Officer doesn't preclude the possibility of help from a more junior official).

If possible, the results should be put together in a report, preferably with maps and photographs, so that copies can be sent to those you are trying to influence — for a start the Chief Executive, Planning Officer, Environmental Health Officer, Chairman of the Planning Committee and councillors for the area concerned at the District Council; the County Planning Officer; the Member of Parliament; the owners of the site; and of course, the press, radio and television. It might be worth discussing the timing of the publication with the news editor of the main local paper to make sure you get maximum coverage. A letter or telephone call will often encourage radio and television to cover it too.

Where land is expected to remain idle for some time there is an opportunity for the voluntary group to make use of it temporarily. Food growing is appropriate, for it can usually be started (and stopped) relatively easily and economically. The waiting list for allotments nationally stands at 117,000 and many would surely prefer a temporary plot to a long wait. More ambitiously, vacant plots can house livestock — an idea with Whitehall backing in the form of a £15,000 grant to Inter-Action to help groups set up urban farms.

On the south bank of the Thames near Tower Bridge in London, an ecological park is being created on waste land with a range of different habitats for nature conservation and field studies. In Covent Garden several wonderfully imaginative gardens have been created on pocket-handkerchief sites

**continued over**

1952
FESTIVAL OF
BRITAIN
AWARD
BEST KEPT BOMBSITE

# HOW TO SAVE DERELICT LAND

**continued**

including a chess garden with tables inlaid with chess boards.

The key to temporary uses is to persuade the owner to grant a licence which need only be a formal letter. The licence should be for at least two years in the first instance, but may be renewed if both parties agree. Sometimes the temporary use may graduate into a permanent one; on the other hand developments sometimes do go ahead as planned and the group must wind up its interim use or transfer to another site.

Local authorities should (some already do) go out of their way to encourage and help voluntary schemes. They can provide help in kind such as skips for rubbish clearance, equipment and technical advice and above all they should be willing (and should announce that they are willing) to grant licences on their own dormant land. They could actually save money in some cases by giving grants. Erecting corrugated fencing around a site of say 30 by 100 yards would cost in the region of £1,800. £180 for seed, timber, plants, etc could enable a group to make a garden. The contrast is so obvious that it has been largely unnoticed."

*Timothy Cantell Vole 8, 1978*

# CONSERVATION OF BUILDINGS

**1975, European Heritage Year, was the big year for conservation. In the first half of that year, in England and Wales alone, permission was given to demolish 182 listed buildings.**

ARCHITECTURAL conservation in this country is now a fact of life. Legislation abounds, even if it is somewhat uncoordinated, and the widespread concern of the public can be measured by the growth of the amenity society movement.

However, much of the legislation turns out to be toothless. Major infringements merit fines of around £250. Frequently, too, the offender and the guardian are one and the same, namely the local authority.

Plans for redevelopment (not as dead a duck as you might imagine) or for long-term road schemes can ride roughshod over nominal protection for buildings which happen to stand in the way. Listed buildings (in a scale Grade 1, Grade 2 *and Grade 2, afforded different degrees of protection accordingly) are supposedly those of architectural and historic merit. Yet many important buildings are not protected and their only chance is that of being "spot listed" when threatened. The DoE has a much depleted band of Inspectors, and at the present rate a proper country-wide survey will not be complete until well into the next century. Meanwhile, as a result of unthinking destruction "sound homes can be lost, small businesses destroyed, areas blighted, scarce resources squandered, and the civilising influence of the past dissipated"

*(SAVE Britain's Heritage, Manifesto).*

**Man is naturally a political animal.** *Aristotle*

HIGH QUALITY HARDCORE
—
EX HEAD-QUARTERS OF LOCAL PRESERVATION SOCIETY

# CONSERVATION OF BUILDINGS

continued

**Demolition is the most obvious threat to important buildings. Neglect is just as insidious, and does not even have the limited protection of "listed building consent" that stands between a listed building and demolition. Dereliction, effectively demolition by default, has been calculated to be threatening perhaps 25,000 to 30,000 listed buildings in Britain. The bad condition of a building was given as a major reason for demolition in almost 2 out of every 3 applications for consent to demolish in 1977.**

MONEY
Government money for historic buildings is administered by the Historic Buildings Council, founded 25 years ago and now with a budget of c.£7m. There are separate councils for England, Scotland and Wales. Grants go to individual buildings of all descriptions, to churches in use (though that is being relaxed to include those under threat), as well as to buildings within Outstanding Conservation Areas — that is an area of town or village of very special significance as a whole. Local authorities have wildly divergent attitudes to the funding of conservation. Chester City Council, with an excellent record, is budgetting £70,500 for its conservation fund in 1979-80, and may allot more. Other county and district authorities produce a derisory thousand or two.

One source of finance is that of the **Building Preservation Trusts;** at the last count there were 38. Some are run by local authorities, some by local individuals or amenity societies. Often they start with a single building and then, on its sale, progress to more with a Rolling Fund. Considerable funds are available to them through the **Architectural Heritage Fund,** administered by the **Civic Trust** and funded from private sources. These loans are then matched by Government money, pound for pound.

REUSE

Many buildings are in a vulnerable position because of economic and social changes. Churches and large country houses are perhaps the most obvious categories. The **Redundant Churches Fund** ("the National Trust of churches") now has 136 outstanding churches vested in its care and adds up to 20 annually. That leaves an enormous number for which the future is either a new use — or demolition. Country houses equally have to be rescued, possibly by adaptation to multiple use, though the alternative for them tends to be dereliction rather than demolition. Public buildings, corn exchanges, town halls, public baths etc. present similar problems. For every happy tale of an imaginative new use there are many more of good buildings lost for want of ideas. One public body which has an exemplary approach to the reuse of old buildings is the **Sports Council** which regards chapels, laundries, market halls etc. as perfect adaptable space, once they meet certain criteria of location, condition and dimensions.

EYES DOWN
CHURCH
HOURS of WORSHIP
Morning Bingo 8.am
EvenBingo 6·30pm

# CONSERVATION OF BUILDINGS

**continued**

BODIES

Apart from the myriad local amenity societies which range from coteries acting out of self-interest to genuinely concerned groups of individuals, there are a number of national amenity societies, four of which, with the Civic Trust, form a Joint Committee — the body which is consulted in cases of extreme concern.

---

Grandfather of them is the **Society for the Protection of Ancient Buildings** founded by William Morris in 1877. With a pre Georgian cut-off date the Society fights for buildings of note, provides technical advice and has a strong line (emanating from its original Manifesto) on the importance of "repair" of buildings rather than their "restoration". SPAB 55 Gt. Ormond St., London WC1

**Ancient Monuments Society**

Founded in the 1920s as the northern equivalent of the SPAB but with a wider remit, interesting itself in buildings of any date. More recently it has been recognised as one of the four national amenity societies, specialising in vernacular architecture and churches. Offers advice on reuse, grant aid and general problems of historic buildings and, like its fellow societies responds to planning applications and is represented at Public Enquiries when necessary.

The Ancient Monuments Society, St. Andrews by the Wardrobe, Queen Victoria St. London EC4.

---

**The Georgian Group** exists to defend the buildings of the 18th century and, like the SPAB and the Victorian Society, will provide evidence at Public Enquiries and give general professional assistance in the event of a threat to an outstanding building.

The Georgian Group, 2 Chester St., London SW 1X 7BB

CHURCH
of
SAINT
BETJEMAN

**The Victorian Society**, the most recent of the "specialised" amenity societies, has perhaps the hardest task. Its function has been largely to educate the public to the virtues of Victorian architecture, much of which had no statutory protection in the 1960s and often has none still — a reflection not of its merits but of the antiquity of many of the lists — as well as fighting for the buildings which are threatened. The change of taste can be measured by the fact that St. Pancras Hotel, once under threat of demolition by British Rail, is now used by that body in publicity emphasising their conservation-conscious outlook. Railway stations are merely one category of building which is, with the exception of a single building, exclusively Victorian or later.

The Victorian Society, 1 Priory Gardens London W 4.

**The Civic Trust**, set up in 1967, coordinates much of the information on conservation with the Civic Trust Newsletter, administers the Architectural Heritage Fund, presents its own awards and keeps a full register of amenity societies, as well as publishing handbooks and reports and offering advice on a wide range of conservation problems.

The Civic Trust 17 Carlton House Terrace London SW 1.

---

**Save Britain's Heritage** is the most recent addition to the list. It does not have a membership, unlike the first four, and its function is that of publicising deserving cases — of buildings of any date and any type. By publishing reports and booklets and by compiling exhibitions it sets out to induce a greater awareness in a wider public of the problems which remain unsolved in the conservation field.

SAVE Britain's Heritage 3 Park Square West, London NW1.

The Historic Buildings Bureau has lists of endangered, listed buildings for sale. 25 Savile Row, London W1.

# CONSERVATION FACTS AND FIGURES

**Churches and Chapels:**
between 1968 and Dec 1978 740 C of E churches & chapels have been declared redundant
272 have found new uses
176 have been or are subject to demolition schemes.
136 vested in the Redundant Churches Fund
(remainder either combination of fates or undecided)
In the last 15 years nearly 1000 churches & chapels have been demolished (all denominations) At least 200 of these were listed.

**Country houses**
in 1974 it was estimated that since 1945 at least 712 notable country houses in Britain had been demolished, gutted or allowed to fall into ruin. Those demolished; 431 in England, 175 Scotland, 23 Wales. "Notable destructions" included that of Deepdene by British Rail in 1969 and of Eaton Hall by the Duke of Westminster, 1961 onwards.

Smaller houses — no record has been kept but Bath has lost an estimated 2000 Georgian houses in recent years. 1 in 10 of its listed buildings remain 'derelict' or in 'bad condition'.

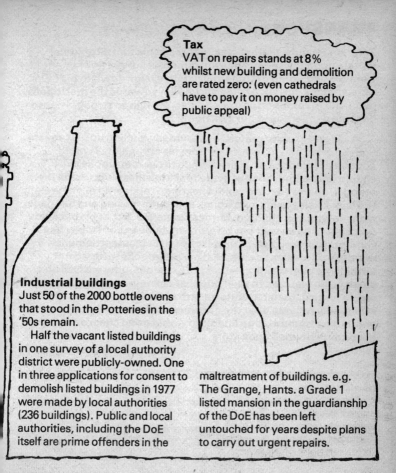

**Tax**

VAT on repairs stands at 8% whilst new building and demolition are rated zero: (even cathedrals have to pay it on money raised by public appeal)

**Industrial buildings**

Just 50 of the 2000 bottle ovens that stood in the Potteries in the '50s remain.

Half the vacant listed buildings in one survey of a local authority district were publicly-owned. One in three applications for consent to demolish listed buildings in 1977 were made by local authorities (236 buildings). Public and local authorities, including the DoE itself are prime offenders in the maltreatment of buildings. e.g. The Grange, Hants. a Grade 1 listed mansion in the guardianship of the DoE has been left untouched for years despite plans to carry out urgent repairs.

# TRANSPORT

Tristram Shandy's father was "very sensible that all political writers upon the subject had unanimously agreed and lamented, from the beginning of Queen Elizabeth's reign down to his own time, that the current of men and money towards the metropolis, upon one frivolous errand or another, — set in so strong, — as to become dangerous to our civil rights . . .

'Was I an absolute prince,' he would say, pulling up his breeches with both his hands, as he rose from his arm-chair, 'I would appoint able judges, at every avenue to our metropolis, who should take cognizance of every fool's business who came there: — and if, upon a fair and candid hearing, it appeared not of weight sufficient to leave his own home, and come up, bag and baggage, with his wife and children, farmers' sons, &c. &c. at his backside, they should be all sent back, from constable to constable, like vagrants as they were, to the place of their legal settlements. By this means I shall take care, that my metropolis tottered not through its own weight; — that the extremes, now wasted and pined in, be restored to their due share of nourishment, and regain, with it, their natural strength and beauty: — I would effectually provide, that the meadows and corn-fields, of my dominions, should laugh and sing; — that good cheer and hospitality flourish once more . . . ."

Tristram Shandy vol. 1, chapter 18

Personal mobility is one of the most basic of freedoms, and is invariably one of the first to be removed by repressive regimes. From this point of view Mr. Shandy's proposal is nothing short of totalitarian. At the same time one can sympathise with his disapproval of the way in which the great cities suck everything towards them, thereby making the cities intolerably overcrowded and at the same time impoverishing the less densely-populated surrounding areas.

Unlimited personal mobility is a fine ideal. The obverse of that bright coin is less attractive, in every way from congested streets and polluted air to sardine-packed commuters and tiny villages jammed (in some cases actually demolished) by juggernauts.

A few years ago there was a view, expounded especially by the late Anthony Crosland that the step in democratic development after "One man, One Vote" was "One Man, One Car". The unpleasant consequences of this reliance on road transport will long be with us. They vary from declining public transport (which for the poor and disabled, especially in rural areas, effectively means personal mobility) to the lousy labour relations which seem to be an integral part of the car manufacturing industry. Every year more and more acres are lost to roads, and more lives are lost on them.

Man is a restless animal, and has never been more numerous or more restless than now. More and more people live further and further from their places of work. Their compensation for 48 weeks of frustrating daily home-work-home travel is all too often relieved by brief breaks and holidays which involve even more horrendous travel experiences — stuck in miles-long traffic jams at bank holiday weekends or castaway in snarled-up airports.

More and more goods are moved around the world or around the country. The growth of large-scale, centralised industries means that manufactured goods are increasingly produced in a centralised manner and then moved to the consumer. People work further from where they live. The clothes they wear, the food they eat, are no longer produced within walking distance but hundreds, even thousands, of miles away.

Presented with this frantic movement of people and goods, it is hard for the voice of common

**continued over**

sense not to ask whether all these journeys are strictly necessary.

One thing that's sure is that it's not very efficient.

The most economical, the most pleasurable and (surprisingly often) the quickest way of getting from A to B is on two feet.

With only a small capital investment, in the form of a bicycle, speeds can be increased and much greater weights moved.

The cyclist needs only one-fifth of the walker's energy to move three or four times as fast.

**"A well-maintained bicycle chain loses only 1.5 per cent to friction, compared with power losses of up to 15 per cent for a car gear-box, which has to turn lubricating oil, idler shafts and all those cogs".** Watson and Gray.

Water is also a cheap and efficient form of transport, both inland and globally. It is decreasingly used. Railways are rapid, safe and efficient. Road transport, by contrast, is expensive, polluting and dangerous. It pollutes the air, it eats up land, it kills thousands yearly, it congests our cities, it spoils our landscapes, and it is

**"A man on a bicycle has the highest efficiency rating among all moving animals and machines. In terms of energy consumed in moving a certain distance as a function of body weight it is far more efficient than a man walking, and it outstrips jet planes, horses and even the sleek salmon. Put it another way: if the energy a cyclist needs from food is compared with the energy from a gallon of petrol, the cyclist is capable of 1,600 miles per gallon."**
Roderick Watson and Martin Gray: The Penguin Book of the Bicycle

overwhelmingly the form of transport that we have chosen.

In 1948 there were 31,000 kilometres of railway in use in Britain. In 1976 there were 19,500 km of which only 15,000 were for passengers.

A cyclist uses 23 kilocalories of energy per passenger mile. A car uses 630.

The energy needed to manufacture one car would be enough to produce between 70 and 100 bicycles.

About 60 per cent of all car journeys to work and 30 per cent of the mileage is for journeys of less than 5 miles. For all car journeys the corresponding figures are 60 per cent and 20 per cent.

The average bus journey in London goes at 3.8 mph — slower than a brisk walk.

Most bus and tube journeys in London are less than four miles in length. Over these distances it's usually quicker on foot or bicycle. And more pleasant. And healthier. And cheaper.

---

ROADS **In 1903 there were 2 cars per thousand population**
**In 1976 there were 257.8 cars per thousand population**
**One person in four now owns a car.**

In 1976 there were 7,985 people killed on roads in Britain. By 1976 this number had dropped to 6,570. Even if one drops this figure to 6,000 and multiplies it by a hundred it is apparent that cars and roads are causers of violent death on a scale easily comparable with world war casualties.

Between 1956 and 1972 there was a loss of 138.5 million bus miles per year, mostly in rural areas.

COUNTRY BUS STOP
LAST BUS – 1957

continued over

# TRANSPORT continued

**Almost a million people have been killed or seriously injured on the roads in the past ten years.**
**Since 1928 325,000 have been killed and 12 million injured. Each year the number killed or seriously injured is about equal to the population of Cambridge.**

---

The motorway network uses up about 25 acres of land per mile. When junctions are included this land-hogging can rise to more than 40 acres per mile. Estimates for the completion of the proposed 2,000 miles motorway network programme suggest that it alone will involve the loss of 22,000 acres of chiefly agricultural land.

For the past twenty years the United States have received an additional 200 miles of paved roads and streets every single day. For the last quarter century the rate of passenger miles in the United States has increased six times faster than the population.

**Ivan Illich's Time Exchange Sum.**

HOW long does a journey take? London to Brighton is 60 miles — an hour by train from city centre to centre; say two hours by car, about 6 hours by bicycle, about 20 on foot. But then you have to add to those figures the amount of time spent in earning the money to pay for the journey. The twenty hours' walk to Brighton merely takes twenty hours. To all the other times must be added the time-money equivalent spent on buying or hiring the vehicle or paying for the railway ticket.

**In 1900 there were 1,115 million rail/passenger journeys.**
**In 1920 there were 2,186 million**
**In 1975 there were 715 million**

The Time Exchange sum can be used for any form of consumption. For example, you can grow your own vegetables or you can buy them. Growing your own involves x hours' work in the garden or allotment. Buying them involves a hours going to the shop and back, plus b hours earning the money to pay for the vegetables, plus c hours earning the money to pay for the journey from home to shop and back. Illich contrasts direct work which we can do ourselves using what he calls "convivial" tools (bicycles, hand-drills, etc) as against indirect work, earning money to pay for unconvivial tools (cars, dish-washers) or to pay professionals to do the work for you.

**❛ The typical American male devotes more than 1,600 hours a year to his car. He sits in it while it goes and while it stands idling. He parks it and searches for it. He earns the money to put down on it and to meet the monthly instalments. He works to pay for petrol, tolls, insurance, taxes and tickets. He spends four of his sixteen waking hours on the road or gathering his resources for it. And this figure does not take into account the time consumed by other activities dictated by transport: time spent in hospitals, traffic courts and garages; time spent watching automobile commercials or attending consumer education meetings to improve the quality of the next buy. The model American puts in 1,600 hours to get 7,500 miles: less than five miles per hour.**

**In countries deprived of a transportation industry, people manage to do the same, walking wherever they want to go, and they allocate only three to eight per cent of their society's time budget to traffic instead of 28 per cent. What distinguishes the traffic in rich countries from the traffic in poor countries is not more mileage per hour of life-time for the majority, but more hours of compulsory consumption of high doses of energy, packaged and unequally distributed by the transportation industry. ❜**

Ivan Illich, *Energy and Equity*

continued over

# TRANSPORT

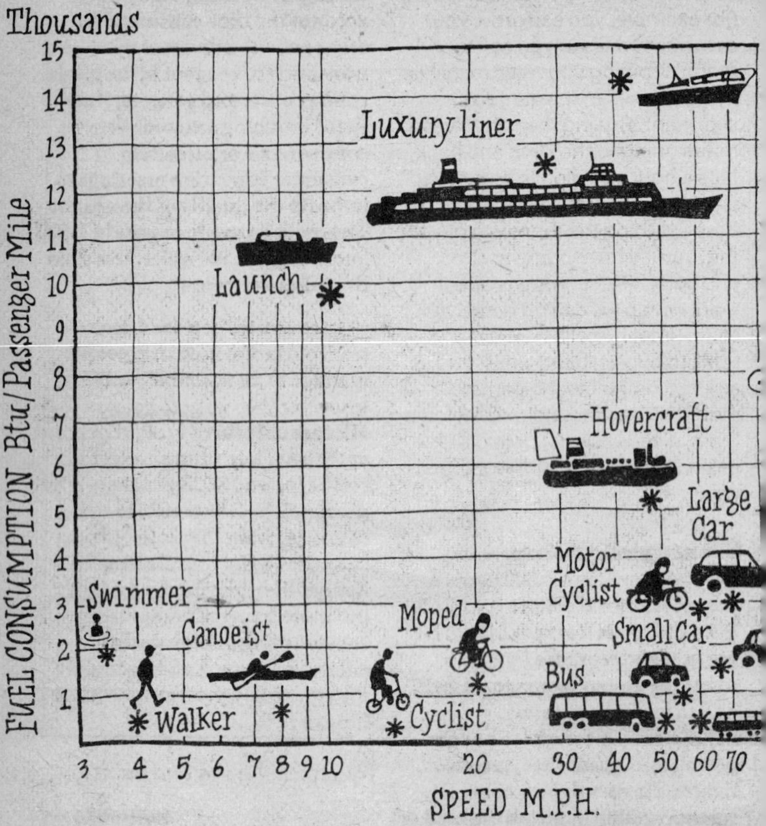
Thousands
FUEL CONSUMPTION Btu/Passenger Mile
15
14
13
12
11
10
9
8
7
6
5
4
3
2
1
Luxury liner
Launch
Hovercraft
Large Car
Motor Cyclist
Swimmer
Moped
Canoeist
Small Car
Bus
Walker
Cyclist
3
4
5
6
7
8
9
10
20
30
40
50
60
70
SPEED M.P.H.

**Different transport modes have very different energy requirements.**

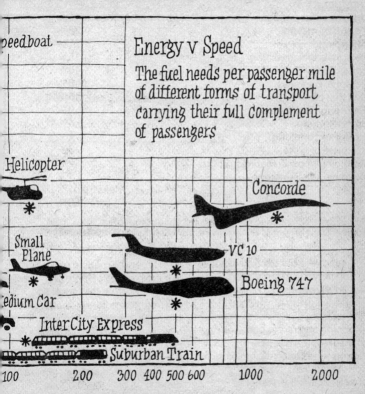

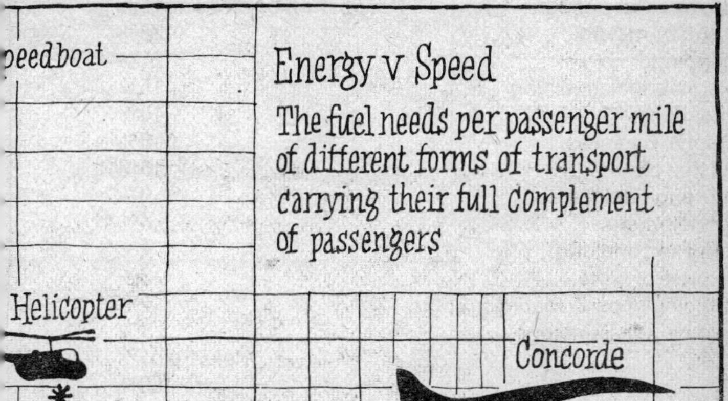
Energy v Speed
The fuel needs per passenger mile of different forms of transport carrying their full complement of passengers
peedboat
Helicopter
Concorde
Small Plane
VC 10
Boeing 747
edium Car
InterCity Express
Suburban Train
100
200
300
400
500
600
1000
2000

Transport 2000; Energy in Transport

**continued over**

Not all journeys are for fun, as the table shows:

| Journey Purpose | Proportion of all Journeys |
|---|---|
| To and from work | 22% |
| Shopping | 21% |
| Personal, social, visiting | 15% |
| Education | 10% |
| Personal Business | 8% |
| Day trip, play | 5% |
| Escorting other person | 5% |
| Entertainment | 4% |
| Eating and drinking | 4% |
| In course of work | 3% |
| Watching or participating in sport | 2% |
| Holidays, pleasure trips | 1% |
| Other | less than 1% |
| | Total 100% |

Note: Figures do not add up to 100% because of rounding.
Source: Department of Environment, 1972/3 National Travel Survey

Access to a car becomes crucial in a society in which public transport is declining, and where people are living further and further from their places of work, shopping and recreation. The new town of Milton Keynes, for example, has been designed for a population density of 8 people per acre, specifically in order to facilitate car use. This constrasts with around 60 people to the acre in a town such as Brighton, designed before the motor age.

Though one person in four owns a car, there are large sections of society that do not have access to one. The young, the old, the poor, the sick add up to a large number. Pensioners, for example, generally depend on walking or public transport.

Car ownership in Great Britain

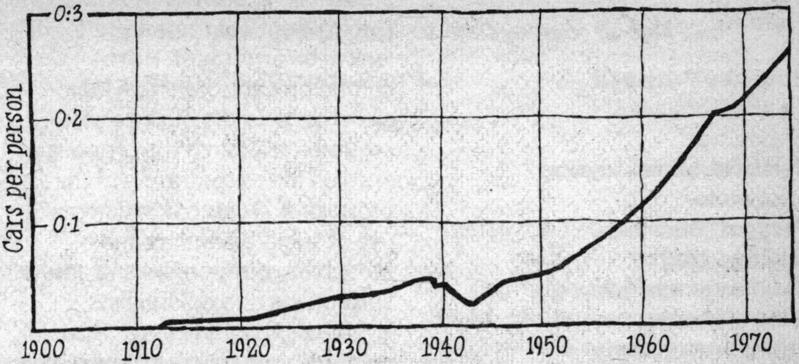

0:3
0·2
0·1
Cars per person
1900
1910
1920
1930
1940
1950
1960
1970


**"For the sole purpose of transporting people, 250 million Americans allocate more fuel than is used by 1,300 million Chinese and Indians for all purposes. Almost all of this fuel is burnt in a rain dance of time-consuming acceleration. Poor countries spend less energy per person, but the percentage of total energy devoted to traffic in Mexico or in Peru is greater than in the USA, and it benefits a smaller percentage of the population."**
Ivan Illich: Energy and Equity

**"The fuel used to carry one passenger on an average trans-Atlantic scheduled flight is equivalent to the anual per capita consumption of energy for all purposes in the Third World."**
Meyer Hillman: Changing Directions.

**The swiftest traveller is he that goes afoot.** *H.D. Thoreau*

**continued over**

# TRANSPORT

**continued**

## ACTION SECTION

IN recent years opposition to the building of motorways has become increasingly vigorous. Some of the campaigns have been extremely successful. The lessons learnt in these campaigns are easily applicable outside the area of transport.

**Enemies** The Department of Transport and the Road Lobby are formidable opponents, but nothing like as great as the real enemies, which are apathy and defeatism. Government still requires the acquiescence of the governed. If enough people refuse to acquiesce, then the proposals cannot succeed.

**Jobs for All** The Department assiduously cultivates the myth that transport planning is the business of "experts". Counter-experts are useful for tearing down the pretentious statistical facade behind which the Department customarily hides its lack of an argument. But a campaign cannot rest on logic alone because the judges of the arguments against the proposals are also the perpetrators of the proposals. A campaign can make use of social scientists and preachers, puppeteers and actors, cartoonists and polemicists, jumble-sale organizers and poster painters, long hairs and short hairs, young and old, and people of all non-violent political persuasions. Every successful campaign discloses truly remarkable reserves of hidden talent in the community.

**Allies.** Transport links everything to everything else. A major transport project will have repercussions far beyond the area where it will be built. Often those who will bear the brunt of these repercussions will be quite unaware of the impending threat. Certainly the Department makes no effort to inform them. If informed and encouraged, such people can make very helpful allies.

**Divided we fall.** The Department is usually quite tolerant of objections to minor details of its proposals, but is extremely hostile to those who query the need for its proposals. If it can provoke the objectors into squabbling among themselves about precisely whose garden or neighbourhood their scheme should obliterate, it will be well pleased. The Department is a staunch adherent to Lord Lugard's dictum "Divide and Rule".

**Time.** With every day that passes the futility of the Department's traffic promoting policies becomes apparent to more people. Time is on our side. The longer that schemes can be delayed the greater the chance that they will never be translated into reality.

All over the world one can now find monuments to the megalomaniac folly of planners in the form of abruptly truncated motorway programmes and abandoned highrise housing schemes.

**Knowledge is power.** Often the Department is extremely secretive about its intentions. It will frequently allude to the existence of consultant's reports or cost-benefit studies which it claims justify its proposals but which it cannot divulge for various flimsy reasons. But on other occasions it will attempt to befuddle the opposition with volumes of undecipherable, poorly referenced information. The Department is typically paranoid in its reaction to requests for information. It believes with some justification that the people requesting it intend to use it against the Department. Keep asking.

**Fundamentals.** It is important to acquire a detailed mastery of the proposals without losing sight of the larger issues at stake. Will the proposals increase or decrease our dependence on cars and scarce energy?

**Policy.** The Department has admitted that it does not have a coherent transport policy. It finds it embarrassing to be reminded of this. Find as many ways as

**continued over**

# TRANSPORT

**continued**

possible to publicise the nature and the consequences of its lack of coherence. Embarrassment is a powerful weapon.

**Education.** One is up against not only the intransigent self-interest of the Road Lobby but also widespread ignorance about the fundamental issues. Fighting a motorway can be an educational experience for all concerned. A public inquiry can be an excellent forum for exploring and publicising the relationships that exist between economic growth, energy consumption, traffic and landuse.

**Job Satisfaction.** Fighting motorways can be very satisfying work. It affords opportunities for sticking pins in balloons and watching bureaucrats slip on banana skins. It is a cooperative venture in which people can take pride and pleasure from their mutual accomplishments. One tends to meet extremely pleasant, concerned, intelligent and creative people among protestors at motorway inquiries.

---

**The public's reaction to our democracy today is not a delight in the feeling that people are mattering more and more in modern Britain, but a growing alienation through the feeling that people are mattering less and less. I believe it is vital that this sense of alienation is removed in the only way it can be removed — by reviving active participation at all levels.**

*Richard Crossman, Granada Lecture*

---

The growth of traffic has created a condition in which children in particular have sustained a dramatic loss of independence as parents have increasingly felt obliged to impose restrictions on them. This has resulted in a considerable circumscription of their freedom as individuals, compared with the freedom enjoyed by their parents and even more by their grandparents. Recent surveys have shown that, on average, parents will only allow their children to cross main roads unaccompanied at the age of eight, to travel by bus at the age of nine and to cycle on the roads on their own at the age of ten, but far from this being recognized as an unfortunate development requiring changes in transport policy in view of this inroad into children's basic freedom, parents are often admonished for not accompanying their young children. Children's inability adequately to perceive traffic danger is viewed as sufficient reason for denying them this freedom: the Department of the Environment specifically excludes them from traffic counts taken to establish the need for pedestrian crossings. It is odd that children are obliged by law to go to school but the law gives them no protection on their way.
Meyer Hillman

**Few things are as immutable as the addiction of political groups to the ideas by which they have once won office.** *J.K. Galbraith*

# ENERGY

IT is the same with energy as with mineral resources and food. The few get most. The 6 per cent of the human population living in North America consume more than one-third of the energy. West Europe takes about 20 per cent, East Europe and the Soviet Union 22per cent, and the rest of the world has to make do with what's left over. One North American uses as much energy as 300 North Africans.

**The release of atom power has changed everything except our way of thinking . . . The solution of this problem lies in the heart of humankind. If only I had known I should have become a watchmaker.**
*Albert Einstein*

**But the conservative, while lauding progress, is ever timid of innovation; his is the hand upheld to caution pause; his is the signal advising slow advance. The word *electricity* now sounds the note of danger.** ***R. L. Stevenson*** **A Plea for Gas Lamps**

## The Generation Game

The most wasteful user of energy in Britain is the Central Electricity Generating Board. Modern power stations turn a bit more than a third of the energy used (mostly in the form of coal) into usable electricity. Most of the rest is wasted, and goes to heat the world's air, sea and rivers.

In the last ten years the efficiency of conventional steam-power stations has improved by about 4 per cent (from an average across all plants of 27.9 per cent to 31.5 per cent).

The system can meet the highest peak demands expected. This is achieved by running for most of the time well below capacity. The average hourly output is about half the peak demand.

With great effort, coal is extracted from the bowels of the earth and taken to a power station. It is burned and 10 per cent of its energy is immediately lost via the stacks as their temperature has to be high enough to prevent the condensation of water. In converting heat into mechanical energy and then electricity a further 60 per cent is lost. Thus a 2000 MW station is losing 4000 MW in steam and hot water. Further losses occur in transporting the electricity from the station to the consumer.

**Electricity is extremely expensive to transport.**

**Relative energy transmission costs over 100 miles**

| | |
|---|---|
| **Oil** | **1** |
| **Gas** | **2½-3** |
| **Coal** | **5** |
| **Electricity** | **16** |
| **Heat (steam, not water)** | **1000** |

**continued over**

# ENERGY

**continued**

Stuart Wilson of Oxford University points out that:

"It is a fundamental law of thermodynamics that the more efficient the efficiency of conversion into power the lower the temperature of the heat rejected, so the present policy of the CEGB of building larger and larger power stations, whether coal, oil, or gas-fired or nuclear-fuelled, results in fewer but larger power stations at a distance from the big cities. They reject heat in enormous quantities at a temperature too low for any large-scale use so far developed."

Dr Wilson regards the time as ripe for a large number of smaller heat engines close to the users so that only one fuel need be distributed, at low cost, and the expensive transmission of heat and electricity minimised.

What Stuart Wilson calls total energy is the joint supply of both electricity and heat by a much larger number of smaller heat engines close to the users so that only one fuel need be distributed, at low cost, and the expensive transmission of electricity and heat is minimised.

**Total Inland Energy Consumption in UK**

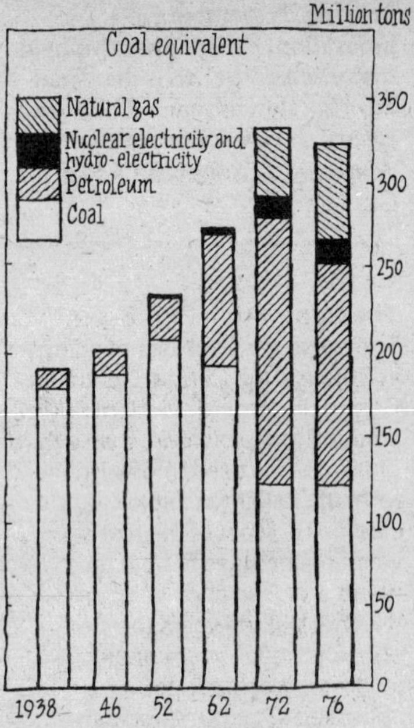

Million tons
Coal equivalent
Natural gas
Nuclear electricity and hydro-electricity
Petroleum
Coal
350
300
250
200
150
100
50
0
1938
46
52
62
72
76

If an all-electric system needs 100 units of energy our present mixed system needs about 75 of units and a total-energy district heating system would need 40 units.

Total energy stations can rise from the usual 35 per cent efficiency to about 80 per cent. In Sweden about a third of the electricity is generated in this way, at a saving of around 78 millions tons of fuel.

COAL at present supplies about one-third of the world's energy. Known coal reserves could keep us going for thousands of year, but it will be increasingly arduous to mine.

OIL production is expected to peak around the end of the century, decline sharply thereafter and run out in the first half of the next century.

**continued over**

**World Oil Production and Consumption 1977**
**million tonnes**

*BP Statistical Review of the World Oil Industry 1977*

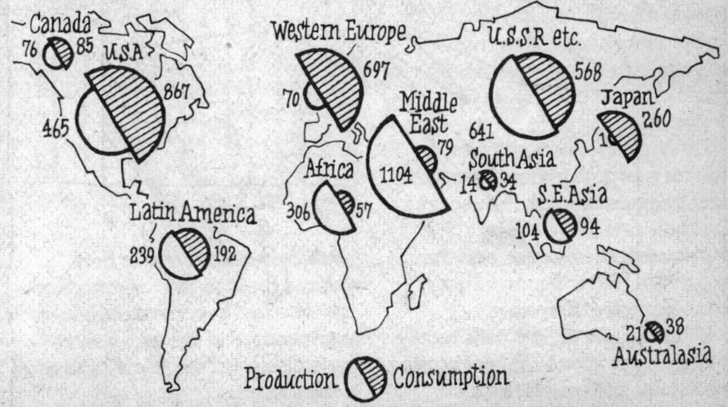
Canada
76
85
U.S.A.
867
465
Latin America
239
192
Western Europe
697
70
Africa
306
57
Middle East
1104
79
U.S.S.R etc.
568
641
South Asia
14
34
Japan
260
S.E.Asia
104
94
21
38
Australasia
Production
Consumption

The page that says it all, from the introductory section of **Nuclear Power for Beginners** by Stephen Croall and Kaianders Sempler, Beginners Books Ltd. 1978.

THIS IS YOUR CAPTAIN SPEAKING...
GOOD AFTERNOON AND WELCOME ABOARD ATOMIC AIRWAYS
WE ONLY HAVE FUEL FOR THE EARLY STAGES OF THIS FLIGHT, BUT WE THINK WE CAN SOLVE THAT EN ROUTE... THE OTHER SAFETY PROBLEMS MAY ALSO BE SORTED OUT IN TIME... OUR TECHNICAL BOYS HAVE BEEN WORKING ON THEM FOR 30 YEARS, AND THEY'LL LET US KNOW IF THEY COME UP WITH ANY ANSWERS...
AS FOR SABOTAGE... WELL, IT'S ALWAYS A POSSIBILITY... AFTER ALL WE LIVE IN DANGEROUS TIMES... BUT YOU'LL HAVE NOTICED THAT YOU'RE ALL HANDCUFFED TO YOUR SEATS AND GAGGED... SO THE CHANCES ARE PRETTY REMOTE... FINALLY I'M SURE YOU'D LIKE TO KNOW OUR DESTINATION...
SO WOULD WE...
ENJOY THE TRIP...
SO LONG, SUCKER...

# ENERGY

**continued**

NUCLEAR power has so far made a small (if dramatic) contribution to energy requirements. There is widespread concern about the health hazards it presents to us and to future generations; about the threat of terrorists getting hold of material to make nuclear weapons; and the threat to civil liberties that such a security-sensitive operation will have (and is already having.)

The most responsible energy policies are those which either use finite fossil fuels as economically as possible or else use non-finite renewable sources such as sun, wind and tide.

**World Primary Energy Consumption million Tonnes Oil/ Equivalent**

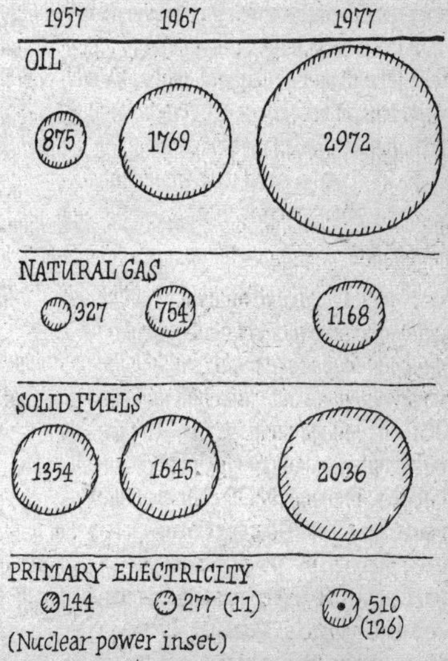
1957
1967
1977
OIL
875
1769
2972
NATURAL GAS
327
754
1168
SOLID FUELS
1354
1645
2036
PRIMARY ELECTRICITY
144
277 (11)
510 (126)
(Nuclear power inset)

# ENERGY

**continued**

## ALTERNATIVE TECHNOLOGY

ENERGY is conserved sun-heat, stored variously, and released for our use variously. We live in an age which uses very old sun-energy (laid down as coal and oil), and uses it up very quickly. The sun is giving us energy every day which we could be using. It is energy which the sun replaces daily. We should learn to think of the technology which harnesses energy in a safe and substantial way as higher, not lower, technology.

About 1.5 quadrillion mega-watt-hours of solar energy arrive at the earth's outer atmosphere each year. This is 28,000 times greater than all the commercial energy used by mankind. Roughly 35 per cent of this energy is reflected back into space. Another 18 per cent is absorbed in the atmosphere and drives the winds. About 47 per cent reaches the earth.

Every year 90,000 billion tons of coal equivalent lands on the earth's surface in the form of solar energy. If only 1 per cent of that could be tapped at about 5 per cent efficiency the whole of the earth's population could achieve the same energy consumption as the United States.

---

**Why assume that the God who presumably created the universe is still running it?** *H.L. Mencken*

---

The UK's sun radiation climate is perhaps half as good as that of the best countries.

Sunlight can be used directly to heat water, or can — by the use of

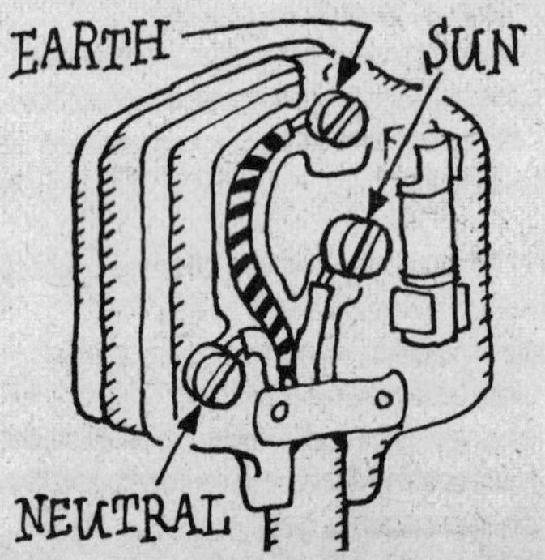
EARTH
SUN
NEUTRAL

photovoltaic cells — generate electricity. The technology of the latter is less economic than the former, at the moment, though is improving. Even Gerald Leach in *A Low Energy Strategy for the United Kingdom*, a book specifically intended not to overplay Alternative Technology in its assumptions, believes that 40 per cent of water heating could be by solar means by the year 2000. Improved heat storage systems will make the variability of solar heat less of a drawback to the system. Preliminary calculations suggest that roughly 34 per cent of end-use energy in the United States is employed as heat at temperatures under 100 C; much of this energy heats buildings and provides hot water. Cheap, unsophisticated collectors can easily provide temperatures up to 100 C.

Like many "new" technologies, solar power was actually developed many years ago, and only abandoned in the energy-glutted 1950s. 50,000 solar heaters in Miami were replaced by cheap natural gas.

## WATER

The world value of hydro-electric power is about 2 per cent of total energy use. It is said that it might be able to rise to about 15 per cent.

At the time of the Domesday book, there were 5,624 watermills in Britain. There were nearer 20,000 by the end of the Industrial Revolution.

## WAVE POWER:

It is said that the UK might get about 50 million megawatt hours per year from a fully developed wave-machine energy system: about a fifth of our current electricity generation, and around 10 times the current hydroelectric capacity.

## TIDES:

The use of tidal variation — especially by allowing a creek to fill at high tide, and then slowly releasing the stored water during the falling tide — is said to be able, if all the likely sites were utilised, to have a potential of about 13,000 megawatts, or about 15 per cent of current total electricity generating capacity.

# ENERGY

**continued**

In India about 68 million tons of cow dung are burnt (very inefficiently) ever year.

30 per cent of India's energy comes from wood. 95 per cent of Tanzania's comes from wood. Reliance on wood for fuel depletes forests and can cause deserts.

In the poorest parts of the world as much as 90 per cent of the fuel is in the form of wood or animal dung.

In 1976, 25,000 excrement converters (making biogas by anaerobic fermentation) were sold in India. There are said to be 2 million operating.

In May 1977 the New China News Agency reported that 4.3 million biogas units were operating, generating methane gas from decaying organic matter.

In the poorest parts of the world people will burn cow dung. This creates heat in the fireplace but deprives the fields of valuable nutrients. According to the FAO figures, about 400 million tons of cow dung are burnt every year in Africa, Asia and the Near East. Each ton of dung burned means the loss of 50 kilograms of potential grain output.

If the methane in the farts of farm animals in the United States could be collected and redistributed, the domestic fuel needs of about 8 per cent of the population would be supplied. This would equal the total home fuel requirements of New York City, Chicago, Los Angeles, Philadelphia and Detroit.

---

**I know why there are so many people who love chopping wood. In this activity one immediately sees results.** *Albert Einstein*

---

The chart is not as complicated as it looks. It shows the use and misuse we make of various forms of fuel.

On the left are inputs. On the right and at the bottom are outputs.

If you look at the input of petroleum, for example, you will see how part goes (after refining) into power stations to make electricity, part goes for transport, part for domestic use, and so on.

More than a quarter of the energy potential available in primary fuels is wasted, mostly by turning coal into electricity. The chart does not account for the further waste that occurs after the electricity has been delivered to the user.

**Flow of Energy in UK, 1972**
**Million Tonnes of coal or coal equivalent**

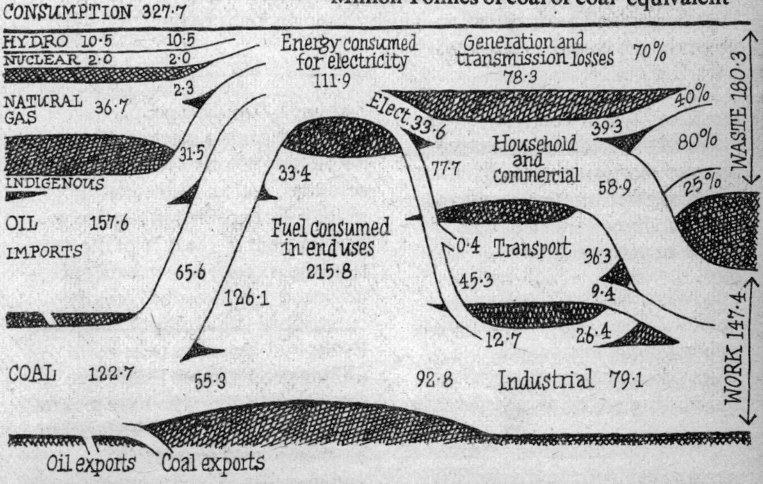

CONSUMPTION 327·7
HYDRO 10·5
10·5
NUCLEAR 2·0
2·0
NATURAL GAS 36·7
2·3
31·5
INDIGENOUS
OIL 157·6
IMPORTS
65·6
126·1
COAL 122·7
55·3
Energy consumed for electricity 111·9
Generation and transmission losses 78·3
70%
Elect. 33·6
33·4
77·7
Fuel consumed in end uses 215·8
40%
39·3
Household and commercial
58·9
80%
25%
WASTE 180·3
0·4
Transport
36·3
45·3
9·4
12·7
26·4
92·8
Industrial 79·1
WORK 147·4
Oil exports
Coal exports

# WORK

THE WORLD faces crucial problems about work. In the richer countries there has been a continuing recession for several years which has meant that for many people there is no work at all. For those in work, there has been a continuing trend toward mindless labour, the requirement for which has been emphasised by an anxiety — especially in Britain — about the perils of low productivity, or the value added by each employee.

There is a language of productivity and economics in general which wholly dominates the political stage. It has to do with "getting the economy going again", and "increasing the size of the cake". The political divisions between the right and left have degenerated to the point where only arguments about the relative distribution of this "national cake" are in question. The orthodoxy does not question the merits of the wealth we create.

Equally, we berate ourselves because each British worker has — in western terms — a rather low capital sum invested in him and makes a rather low return on what equipment he has. Discussion of whether the high-technology, high-capital, high-energy goods we make by the high-technology, high-capital, high-energy work-means which now dominate our lives is heretical. But very necessary.

About 200 million people have flooded the labor markets of developing countries during the seventies, and an additional 700 million are expected to require employment by the turn of the century. Already, the number of prospective workers has greatly outstripped the supply of new jobs. By the mid-seventies nearly 300 million people, more than t ree times the number who have jobs in the United States, were believed to be unemployed or severely underemployed, eking out a precarious existence as casual laborers, street peddlers, shoe-shine boys, and other fringe workers.

More than 30 million jobs must be created each year over the next 20 years merely to keep pace with expansion in the Third World's labor force. Anything less is likely to aggravate inequities and lead to rising levels of poverty. If at the same time productive employment is provided for those who are now grossly underemployed — a critical dimension of any effort to lift the incomes of the poorest people — about one billion new jobs must be created by the year 2000.

In many poor countries, entrants into the job market outnumber new jobs by two to one. The work force in India was projected to increase from 210 million to 273 million during the seventies. There are 100,000 new entrants to the Indian labor force each week. At least 15 per cent of the labor force is already unemployed; another sizable percentage is underemployed. Data for scores of other countries now show the same trend. ❜

*Worldwatch 2*

If anything, these figures make the mistake of assuming — if only in the language they use — that what a person who needs a living wants is a job: "jobs" are things, structured, organised employment. One "provides" jobs. A more constructive approach is merely to think in terms of people being able to exercise a skill. Making it possible for people to grow a proportion of their own diet is by no means the same as giving them a job, but it may be a far richer thing for them to have.

Agricultural work has for centuries been cruelly hard. Bad as it was, farm workers were amongst the first luddites. They could see that the mechanisation of farm labour was not being introduced to make their lives easier — as it could perfectly well have been — but to displace them as rapidly as possible. Technology has in general not been used primarily to make the life of a working community easier, but to remove the most troublesome

**continued over**

# WORK

**continued**

factor of production, people, and to substitute energy and machines for them.

In 1920 United States agriculture employed around 25 billion man hours of work. By 1970 this had shrunk to 6 billion.

In the UK, the number of farm workers has pretty well halved since the war, though the outflow of workers has now fallen to about 3 per cent per year.

---

The Way We Work

CAPITALISM'S greatest hero, Adam Smith, believed that the ideas he was propounding amounted to "the natural system of perfect liberty and justice". Very much like Kropotkin later, he envisaged a society of small producers. It was the essence of his ideas that no one individual or clutch of individuals could influence the market by any form of monopoly.

---

**The more ample the size and functions of the modern State, the less opportunity has the average citizen to take an important part in the disposal of its business.** *Harold Laski*

---

Kropotkin celebrated the new possibilities of modern technology. He noted with pleasure how a machine like the bicycle could be made in small factories — workshops — each with its own highly-independent source of power. Highly flexible power, localised means of production, scattered production, individual, autonomous units of production, were his ideal. Kropotkin believed they would arise in every industry in the country, from agriculture to the production of bikes. The exact reverse of this dream has come true.

The number of agricultural holdings has halved in the last century and the size of the average holding has rocketed. In bike manufacture, the number of workshops actually building machines has shrunk from around 746 to less than 50 with a preponderance of the market in the hands of one firm, Raleigh.

Curiously, both capitalism's successes and failures equally lead to centralisation. Success, because it makes firms powerful enough to buy out others. Failure, because it forces governments to conduct centralising mergers of lame-ducks, as in the car industry.

It's true that some miserable jobs have been lost and we need hardly mourn them. Yet many of the jobs of the past could be recreated today in what Ivan Illich would call a "convivial" (humanistically oriented) way by the use of modern technology. Small, high-powered handtools, for instance, could nowadays put Kropotkin's workshop ideal into practice. Modern hydraulic equipment could put many previously back-breaking jobs in the class of the mildly athletic.

The craft industries have taken a terrific beating from rises in property costs, which have turned our city centres into bureaucratic deserts, and deprived them of workshops. Sex has replaced tailoring as Soho's major industry, in one small exception to the rule.

It is estimated that 3-4,000 workshops have been lost in Clerkenwell (half a London borough) in the last 50 years.

School-leavers, now kept at school for an extra year until they are 16, are taught by teachers who have never left school. Youngsters are taught to be cerebral rather than manual and creative. They have been trained to compete for bureaucratic or mechanised jobs, and therefore the language of craft workshops of any kind is alien to them. The few crafts that survive nowadays are starved of youngsters to learn them.

Schools could release children at 13 or so for pleasant, light, creative, disciplined work with craftsmen for as many days a week as the children are contributing to the workshop. They'd learn patience as well as many skills.

**continued over**

# WORK

**continued**

## Frederick Winslow Taylor and Kalmar

BOTH Lenin and Henry Ford commended the work of Frederick Winslow Taylor, the father of Time and Motion studies. "In the past", he wrote, "The man has been first. In future, the system must be first. All possible brain work should be removed from the shop. The time during which the man stops to think is part of the time that he is not productive".

On these principles Henry Ford succeeded in increasing car output per man about 100 times over that of the older craft garages.

At Kalmar, Sweden, Volvo are trying a different system altogether

❛ Cars are assembled on individual platforms that move from one work area to another where teams of workers perform a series of tasks. Thanks to buffer zones, between work teams, where auto bodies can be temporarily stocked, mechanics can vary the pace of their work and take a break without upsetting the next team. Workers help each other and seldom do the same job in succession.

The results of the Kalmar experience have been encouraging. It does not require more time than usual to assemble a car at Kalmar. During the first half of 1976, absenteeism was one-fourth lower than at a conventional Volvo plant of the same size, and job turnover was one-fifth lower. Despite initial construction costs that were about 10 per cent higher than for a normal plant, both labor and management representatives think the advantages of the new system offset the extra cost. ❜

Worldwatch 25

---

**There are 350 million unemployed or underemployed people in the world. Over the next generation 900 million more people will enter the global labour force.**

---

## How Much Does a Job Cost?

In general, it can be argued that high capital investment in a job will create high productivity at the expense of quality. Only those investments which make work pleasanter or the product better need be contemplated. (This includes a similar re-learning job to that of youngsters seeing the pleasure in craftwork: consumers must learn to see the pleasure of quality rather than quantity in what they consume.)

Modern jobs are very expensive to create.

Capital investment per employee USA 1977 (dollars)

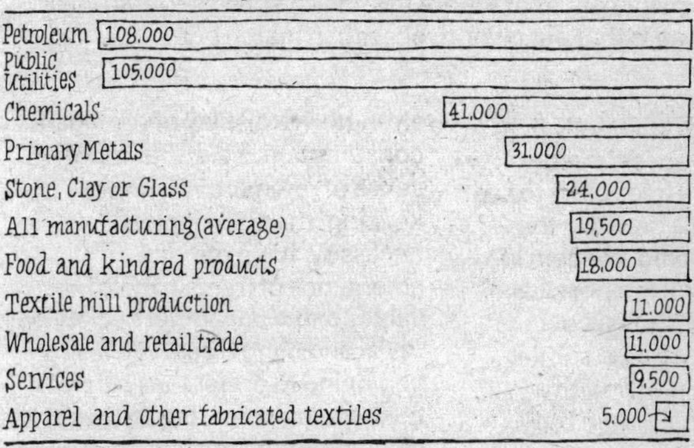
Petroleum 108.000
Public Utilities 105.000
Chemicals 41.000
Primary Metals 31.000
Stone, Clay or Glass 24.000
All manufacturing (average) 19.500
Food and kindred products 18.000
Textile mill production 11.000
Wholesale and retail trade 11.000
Services 9.500
Apparel and other fabricated textiles 5.000

It now costs an average $20,000 to establish a single workplace in the United States, and industrial jobs in the Third World are no cheaper.

Governments will have to ease planning, education and taxation rules before alternative, workshop-based industry can re-establish itself. The re-emergence of new industries will provide a great deal of work, but it will be at low profit, and with small cash flows: rents, wages and taxation will have to be low to create the right climate. Instead, the trend has been steeply upward in all these.

# WASTE

THE twentieth century could be called The Age of Waste. The real proof that something has been consumed is that when we've had the use of it, we throw it away, or at least we throw away what we don't want, or what we can't quickly use. Our attitude to many things, from cars to newspapers, is like that of naughty children to food. We take one bite, and reject the rest.

Part of the problem is that for many years now we have been taught that this is a rich planet and that we are the richest species ever to walk upon it. The notion is that our inventiveness can match our extravagance.

Clearly, we are colossally rich, and our inventiveness is very great. But we are beginning to see that it may be fully taxed in the process of providing a decent life for everyone. That is a physical constraint. But we are also beginning to sense a far subtler constraint to do rather with a sense of properness than the absolute and physical constraints there undoubtedly are.

Many people are beginning to be outraged by the casual and arrogant processes of the consumer society by which each person in the rich world is encouraged to display his or her economic and psychological strength in a display of profligate consumption. This is far more a sense of re-emerging 'house-keeping' than it is of physical necessity. It is a re-emergence of a sense that the values of the consumer society are trivial and dangerous. We are beginning to prefer the notion of husbanding resources to that of squandering them. We are beginning to prefer the idea of leaving behind us greater real wealth than a diminished wealth. The notion that renewal is preferable to depletion begins to impress itself upon us.

Besides, 'consumption' and waste are part of an economic

process which keep millions chained to factories. An economic colossus has been built on endlessly increasing consumption and production. We are beginning to prefer something 'built to last' and to want to spend our working lives making such things by means appropriate to high quality rather than making shoddy goods by the factory means appropriate to disposability.

## HOUSEHOLD WASTE

BETWEEN 17-20 million tonnes of household waste are picked up by local authorities every year in Britain. That works out at something like 2 lbs of rubbish a day for each of us.

Each of us throws away about 5 times our own weight in bottles, jars, tin cans, newspapers, old rags, plastic cartons and leftover food every year.

Even though there has been hardly any cinder content in our rubbish since the Clean Air Act of 1956, there is nowadays a renewed tendency to increases in yearly wastages per capita, after a drop between 1935 and 1968

The picture by volume, as opposed to by weight, has changed very dramatically: in place of heavy cinders, we now throw away vast quantities of packaging. The volume of rubbish has risen perhaps 50 per cent since the Thirties.

In America, the per capita generation of household rubbish is something like 5 lbs per day.

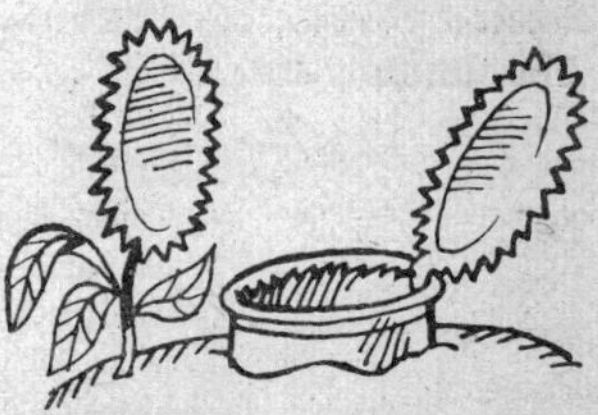

continued over

# PAPER

BETWEEN 35 and 40 per cent of our municipal rubbish is paper, about 10 lbs per average family in the UK per week. In 1974 the UK used about 8 million tonnes of paper, each tonne representing between 15 and 20 trees. In that year we imported about 2 ¼ million tonnes of wood pulp (a tonne of wood pulp makes about one third its weight of paper) and about 3 ¾ million tonnes of paper and board.

About 30 per cent of paper consumed is recycled: in the last war the figure reached 60 per cent.

Recycled paper has many other possibilities than recycling as paper: it can be burned and can produce alcohol.

There are notorious difficulties in the domestic recycling of paper. Recently, many boroughs have grown disenchanted with the erratic economics of their recycling plants (some of them merely baled the waste for recycling elsewhere). The world price of paper and wood pulp is not fixed on ecologically sound grounds, but on what is possible with wasteful and dangerous forestry techniques.

**Proportion of Britain's Raw Materials Used In Motor Car Production.**

| | |
|---|---|
| Steel | 20% |
| Aluminium | 10% |
| Copper | 7% |
| Nickel | 13% |
| Zinc | 35% |
| Lead | 50% |
| Rubber | 60% |
| Plastics | 5% |

# SCRAP METAL

THE UK relies on imports for most of its iron ore and a large part of its non-ferrous metals. Scrap contributes about half of our steel output, and the aluminium, copper, lead and zinc industries all use significant quantities.

The average car weighs about 2,000 lbs, most of it steel, but there are considerable quantities of zinc, copper and lead. We scrap about a million cars a year, and they represent a considerable scrap-source, of which too little - it is said about a quarter - is properly recycled.

Recycling of scrap is far more energy efficient than the smelting of it: for iron: but the price of scrap is often high because of transport costs. Too little account is taken of the environmental costs of original production as against recycling.

# PACKAGING

ABOUT 7 million tonnes of packaging is thrown away each year in the UK. It constitutes about 40 per cent of domestic waste, and is rising. Though most packaging materials are rising yearly, the greatest rise has been in the use of plastics.

Between 1963 and 1971 the per capita consumption of food in the USA rose by 2.3 per cent. In the same period the tonnage of food packaging rose by 33.3 per cent.

USA packing accounts for 75 per cent of all glass, 40 per cent of paper, 29 per cent of plastic, 14 per cent of aluminium, and 8 per cent of steel production

The average UK family throws away about 12 cans a week, whose scrap value is rather less than 1 new penny. Though we lose around 1 million tonnes of steel a year because of this wasteful form of packaging, the expense of detinning the cans makes the steel useless for scrap purposes.

The UK manufactures about 7000 million glass containers yearly in at least 1500 shapes. We throw away 80 % of them after one use only. We produce about 6000 million steel and aluminium cans a year.

RUST
IN
PEACE

continued over

# GLASS

Glass is a relatively cheap material. Its raw materials are cheaper than those for any other packaging form. Its energy cost in manufacture is similar to those for tin cans or plastic bottles.

It is eminently re-usable: and this means that it can be made into bottles which can have a long life, being returned to the distributor for re-filling. Recycling, on the other hand, means that the container's material is broken down and remade as a fresh container.

But part of the problem of modern distribution systems is that they involve a great deal of transport: the manufacturers of drinks, for instance, are a long way from their markets. They don't want the expense of returning empties over long distances. Moreover, supermarkets, which now dominate the retailing scene, are not run with the staff levels which could manage sorting and return of empties.

But above all, there is a vast variety of bottles in use, and so they are not interchangeable between manufacturers and uses.

These are all social, and not at all physical, constraints on sensible packaging.

The Glass Manufacturers' Federation has been promoting a scheme called 'Bottle Banks', which are skips placed strategically in certain towns. The idea is that this is a conservationist move by the industry. In fact, it merely neatly dodges crucial and difficult issues whilst paying lip service to ecologically sound principles and keeping glass manufacture high. It actually takes hardly less energy to recycle glass than it does to manufacture it in the first place.

It would be a simple matter to impose standardisation of glass bottles and to arrange a deposit system which made it attractive to reintroduce a reusing system.

# A BETTER SORT OF POLICY

**(And a Different One.)**

IN the USA Oregon has since 1972 imposed a deposit system on all containers (whether glass or can) of beer and carbonated drinks, and banned altogether the ring-pull opening system. Standardised bottles suitable for re-use carry a lower deposit.

Sweden and Norway have taxation and deposit schemes for beverage containers. Denmark prohibits the sale of non-returnable soft drink containers.

In Sweden, anyone buying a car must pay a deposit: the payment is more than reimbursed when the car is scrapped. If the final owner of the car does not de-register it (at the same time receiving the reimbursement) he or she is presumed to continue to own the car and must go on paying taxes on it. This is to outlaw dumping.

In poor countries there is of course an in-built recycling process. The poor utilise the rubbish of the rich. This has achieved a very high form in Cairo, where — at least in middle class areas — small boys, organised by protectors collect domestic refuse and take it to markets for resale, or piggeries, or for composting for agriculture. Unfortunately, the poorer areas don't provide such rich pickings.

**continued over**

# FOOD WASTE

IT is said that about 20 or 25 per cent of food is lost in the UK in varying forms of waste in storage, distribution and use.

Overconsumption is a form of waste, and the Ministry of Agriculture and Fisheries estimate that the average person in the UK buys 35 per cent more protein and 14 per cent more calories than are actually useful to him or her.

There isn't a great deal of modern data on UK kitchen and plate waste (most of the research being done during the war), there is some fairly modern American data, quoted in FOE's Earth Resources Research's Wastage in the UK Food System. (p.26)

# DUMPING: THE ONLY WAY?

THE majority of our waste is dumped in holes in the ground at an increasing cost to the environment in direct land-use and transport costs both financial and social. Very little of our domestic waste is incinerated, and much of the incineration is not done with any utilisation of the heat created.

This represents a colossal waste of potential power. Municipal waste has a potential heat value roughly a third that of coal.

It is said that there is an additional resource of about 65,000 tonnes of waste oil available, which at the moment seems most useful as a heat fuel.

We seem to be wasting a heat resource worth about £200 million, and that takes no account of the environmental advantages of not dumping the waste from which it comes.

There is some difficulty at the moment in using waste for heat. But the technology is improving and there are several plants in the UK which are developing it.

**Centre for Alternative Technology, Machylnlleth, Powys, Wales.**

A registered charity, sponsored by the Society for Environmental Improvement, the centre is a working demonstration, independent of mains services, of ways of living which use only a small share of the earth's resources and create a minimum of pollution and waste. Has solar collectors, windmills, organic vegetable growing. Open every day of the week (Feb. to Nov.) from 10am to 5pm. Closed December and January. Admission charges: adults 60p, children under 16, OAP's, and students 30p.

Publications: wide range covering DIY, AT, crafts, and environment.

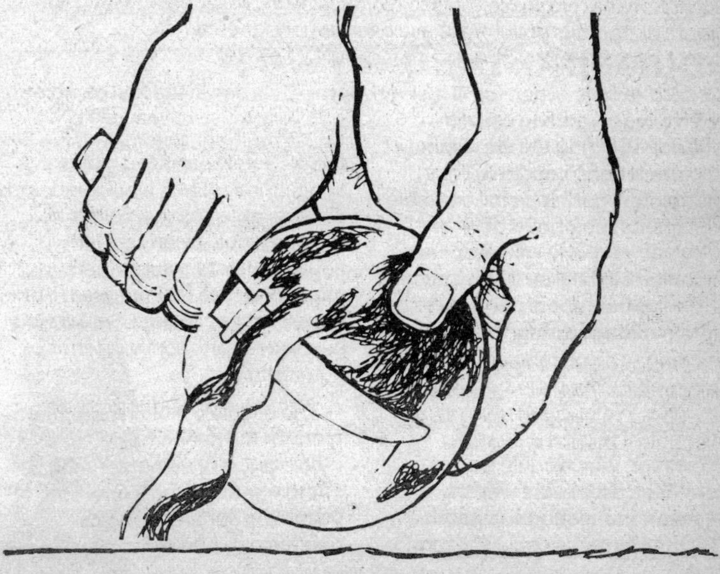

# POLLUTION

"There are several pollutants that are more worrisome today than they were before the birth of the environmental movement," declared Denis Hayes, the founder of Earth Day, in a new Worldwatch Institute study published in Washington in March, 1979. "The buildup of these pollutants — atmospheric carbon dioxide, toxic substances, and nuclear waste, to name only three — poses threats too grave to be ignored."

"Belching smokestacks are much harder to find now than they were a decade ago. However, the pollution problem did not end when they disappeared. Scrubbers and filters have controlled visible forms of pollution that are susceptible to technical fixes. But longer-lasting and more dangerous pollutants have been largely ignored."

For example, when fossil fuels are burned, carbon in the fuel combines with oxygen, adding carbon dioxide ($CO_2$) to the atmosphere If current consumption trends of petroleum, gas, and coal continue the preindustrial level of atmospheric $CO_2$ will have doubled by the year 2020. This $CO_2$-ladened atmosphere will retain more of the earth's heat, leading to rising temperatures. The net global effect is impossible to predict. Widespread melting of the polar ice caps is possible, however, which could lead to dramatic rises in the world's oceans, inundating coastal cities. Rainfall patterns could shift and the delicately balanced global agricultural system would undergo considerable change. Some regions would clearly suffer adverse effects, while others might find their lot improved. But the process of change itself would be tortuous and costly in terms of human life.

The deadly pollution potential of many of the dangerous by-products of manufacturing is another problem only now becoming apparent, Hayes pointed out. The U.S.

Environmental Protection Agency estimates there are 638 chemical dumps in the United States that pose "significant imminent hazard to human health." Many metals, such as mercury, lead, and nickel, are harmful when they are inhaled or ingested. The legacy of these wastes is apparent — at Minamata, Japan, where more than 1,000 people have been deformed by mercury pollution, and in Love Canal, New York, where a whole community built on a chemical dump had to be evacuated.

The problems posed by toxic substances are serious. The benefits derived from some of them are similarly great. The most successful way to control toxic substances would be to keep them in circulation as useful products rather than to discharge them into the general environment. For example, in the United States, it is estimated that more than 70 per cent of the mercury used each year escapes into the environment. If products are designed to be reused and recycled instead of thrown away, less metal would be used overall, and less of what was used would be released into the environment.

A similar long-term pollution problem has emerged with the creation of nuclear wastes. No country has yet found a permanent solution to the problems connected with the by-products of nuclear power. Greater degrees of safety can always be provided for the wastes at greater costs, but absolute and timeless safety can never be assured. Some of the wastes — notably fissionable isotopes of plutonium and uranium — can be recycied, with the formidable danger that some may be diverted into weapons production. Other radioactive wastes, however, can only be isolated from human society for a very long time. Their safety will require a degree of international social and political stability unparalleled in human history.

Long-lived pollutants — such as $CO_2$, toxic substances, and nuclear wastes — can pose

**continued over**

# POLLUTION

**continued**

dangers for thousands of years, or even forever. The ill effects to be felt in the distant future are often severely discounted or even ignored by analysts who make decisions with only the short-term outcome in mind. The problems raised by these pollutants are a challenge to our social values and institutions, to our capacity to forswear a course of development that provides clear benefits in the short run but unacceptable costs in the future.

There are encouraging signs that the general public may be closer to the heart of the issue than are the experts. The swift U.S. response to the chloro-fluorocarbon threat to the ozone layer, the passage of the U.S. Toxic Substances Control Act and Japan's Compensation of Pollution-Related Health Damage Law, and the mounting opposition to nuclear power suggest people are willing to make some basic changes in their life-styles and common business practices when the danger from pollution is apparent.

Whether change will come soon enough is yet to be seen. When pollutant effects are cumulative, time is of the essence. Delay may lead to irreversible damage. Sooner or later the neglected dimensions of the pollution problem will have to be addressed.

Denis Hayes *Worldwatch*
March 1979

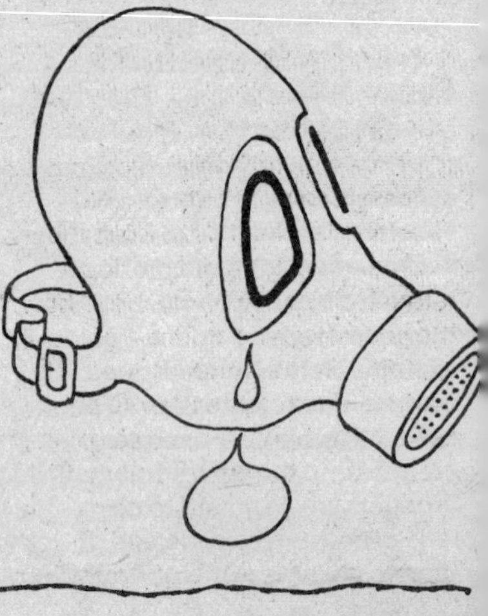

AT the beginning of 1972 the Yorkshire River Authority was in control of more than 296 miles of grossly polluted rivers, which were deteriorating still further at a time when rivers in the rest of Britain were improving. More than 70 per cent of discharged effluent was untreated. A major campaign was sparked off in six articles by Geoffrey Lean in the Yorkshire Post in April and May 1972, naming 25 firms and local authorities as polluters. The Yorkshire Post also found that the Yorkshire River Authority was failing to do its work because of a minuscule budget, and very rarely used its powers of prosecution. The public response was tremendous. Local people took it upon themselves to monitor the levels of contamination and supplied local MPs with a steady stream of reports on the effects on wildlife. Schoolchildren were taught how to measure pollution in terms of biological oxygen demand. Anglers rescued blind and fungoid fish from the rivers and treated them for pollution-caused diseases in their own homes.

Within three weeks of publication of the articles the River Authority took drastic steps to clean up. They increased the pollution prevention department by 250 per cent, gave them new offices and a larger laboratory which would handle 20,000 samples per year. The old laboratory had dealt with 4000 samples annually. Within one year of the start of the campaign all the firms and local authorities named had taken action to clean up the discharges to the rivers. In the two years immediately following the campaign the River Authority made 35 prosecutions on pollution charges. In the three years prior to the public outcry they had prosecuted on only eight occasions.

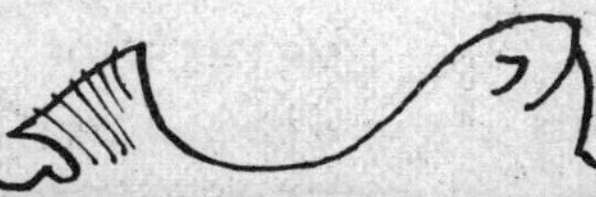

# INSULATION

## BLOCK UP DRAUGHTS.

1. Hold your hand to door and window frames, gaps in floorboards.

2. Fill gaps with papier mache. Stick-on foam insulation strip goes round window and door frames. Or tack metal weatherstrip round.

3. Hang a thick curtain over the front door.

4. Make a long bolster to push against the bottom of the door.

5. Seal up waste outlets in bathroom (at least while bathing), air vents in the larder (at least during winter).

INSULATION Special Offer
BUY NOW
FREE ADVICE
SPECIAL HEATING SUPPLEMENT
SAVE IT 10% OFF

## INSULATE WALLS

1. Cavity walls (homes built after 1920 only): get a reliable professional firm to squirt foam or rockwool into the cavity.

2. No cavity? Put a lean-to shed on the windy side of the house and a conservatory on the other.

3. Insulate inside walls with heat proof Gyproc or fibreglass wallboard, fixed directly onto walls or between battens. Not suitable for damp areas.

Or

4. Line walls with polyurethane or expanded polystyrene. Plastics produce a poisonous black smoke if they catch fire. You must cover them with pasteboard for safety.

5. Line northern walls with cupboards; other walls with bookshelves closely packed with books.

## LINE THE ROOF

1. If the roof is flat hire a professional builder. Insulation of

an existing flat roof will be expensive. If you are in the process of building a flat roofed extension, get the builder to insulate at the same time.

2 Attic: lay down rolls of quilting or glass fibre.

Or

3. Scatter 3in or 4in of granulated vermiculate or mineral wool pellets onto the floor.

NB Tread on joists or your foot will go through to the room below.

## DOUBLE GLAZING

An expensive process. Worth the expense if you have a lot of glass. It might be worth treating just the living room.
There are three main types:
Factory sealed units probably the most satisfactory, certainly most expensive.
Secondary windows: custom built to fit existing windows.
DIY. Several plastic or wooden framing kits available to fit the new panes into. Will have to be removed in summer or if you want to open window.

## PIPES AND CISTERNS

Lag the hot water tank. You can get jackets for this in DIY shops. An old eiderdown will work too. Tie it tightly round so complete tank is covered.
Circular foam lengths are available for water pipes. Two sizes available. From DIY shops or builders' merchants. Lag outside pipes too.

## MORE INFORMATION

You can get grants for roof insulation. Ask at your local Town Hall. The National Home Improvement Council, 26 Store Street, London WC1E 7BJ can advise on insulation.

Glass and Glazing Federation 6 Mount Row, London W1Y 6DY, advice on double and safety glazing.

Draughtproofing Advisory Service, PO Box 305, Bushey, Herts WD2 3HF. National Cavity Insulation Association, 178-202 Great Portland Street, London W1N 6AQ.

# HERB GARDEN

A well-planned herb garden can look decoratively Elizabethan and will provide delicate and unusual flavours to all cooking. If you have no garden, grow your herbs in pots and window boxes. Here are some herbs no garden or kitchen should be without:

***Parsley***. Bi-ennial but often grown as an annual. Best when fresh. Will germinate quicker in warm damp soil and darkness (try the linen cupboard). Scatter chopped parsley over soups and other hot dishes just before serving. The leaf stalks can be chipped with the leaves for flavouring soups.

***Basil***. Delicate annual. Likes warm, sunny position. Grow from seed. Germination should take 10 to 14 days. Will not succeed out of doors in cold, wet summers. Pinch out tops. Water regularly.
Use freshly chopped on tomato salads and sauces, also green beans and aubergines. Drying is difficult, so use only fresh.

***Chives***. Hardy perennial with a subtle onion flavour. They need nitrogen and potassium and chalk. If there is lack of food in the soil the tops will turn yellow. Divide every third spring and dress with compost in autumn. Marvellous chopped up in cream cheese, as a garnish for soups and in green or tomato salads.

***Sweet Marjoram*** (Oregano) Hardy perennial. Balmy scent and flavour, marvellous with Pizza, milanese sauces and tomatoes, potato soup, rabbit stew. Use leaves only, crushed or chopped.
Likes warmth and a neutral soil. When sowing, mix the tiny seeds with wood ash and sow in shallow drills. Should have shade until well established. Water well.

***Mint*** Hardy perennial of innumerable varieties eg spearmint, apple mint, pineapple mint etc. Spreads like mad. Cut during the growing season when the dew has dried. Use for mint jelly and sauce, in fruit drinks, in salads and cream cheeses and in the water when cooking peas.

***Further Reading***

***Herb Gardening*** by Claire Loewenfeld (Faber 1964);
***Herbs from the garden to the cooking pot*** by Robert Quinche and Eugene Bossard (Foulsham, 1975).

# CROPS

## INDOOR CROPS

No garden? Grow your veg in the house.
You can grow tomatoes, cucumbers, courgettes, green peppers, alfalfa and mung beans, mustard and cress, lettuce, even strawberries and melons. Offices often have the sun catching glass and windowsills right for growing.
Indoor vegetables need plenty of light, water, humidity and constant temperatures. They don't like draughts.
Sling shelves across windows with ugly views and grow your plants on these.
Bathrooms are often light and moist, just right for some plants.

How to grow:

***Mustard and Cress***: put a wad of tissues or flannel or J cloth in a saucer or pie dish. Keep it moist. Sprinkle seeds on surface. Keep in dark place until strong growth starts. Eat in about two weeks. Cress takes a day or two longer than mustard.

**continued over**

# INDOOR CROPS

**continued**

***Mung Beans***: Sprout as for mustard and cress but soak them first overnight. Put growing tray into polythene bag and keep in warm, dark place. Eat when about 1 ½ in long.

***Alfalfa***: as for mustard and cress or in a jar. Wash seeds, put four teaspoonfuls in jar. Half fill with tepid water, shake, drain and cover top with muslin fixed with elastic band. Keep on side in warm place (drawer for instance). Three times a day half fill jar with water and drain. Eat in about five days. Rinse first.

## OUTDOOR CROPS

You don't need much garden to grow a lot of food.
Salad foods are usually quickest and easiest to grow and only take up a little space.
Radishes can be mixed with seeds that take longer to germinate so you can see where the rows are. When the other seeds appear, eat the radishes.
Small varieties of lettuce can be grown in pots, or grown among flowers in a herbaceous border — a traditional country custom.
Grow tomatoes in pots or growbags against a south facing wall in a protected situation.

If you have a vegetable plot dig it well. Thrust the spade in vertically — not slanting. To dig one spit deep take out a trench of soil the width of the plot. Turn the soil from the next strip into the trench and so on.

You will need to plan your crops to make full use of the space. Feed the soil well with well-rotted animal dung or compost. Liquid manure is useful during the growing season. Rake in some fertilizer a few days before sewing or planting. Divide the plot into three and rotate in this order: legumes, roots, brassicas.

***Easy to grow vegetables include:***

Runner Beans: sow from mid-May until end June in rich soil in a double row 2in apart. Remove alternate seedlings to thin. Mulch with strawy manure. Water thoroughly, pinch out shoots when plants reach top of stakes. Pick regularly while still small. Cook until just tender. Eat hot with butter or as salad. Runner beans freeze well.

Beetroot: sow from mid spring to early summer. Drills should be 1 inch deep and 1 ft apart. Water and hoe well. Protect from birds and snails. Harvest while still small. Cook in skin or the colour will bleed. Boil or steam and eat hot or cold or use in borscht. The cooked tops can be eaten like spinach.

Spinach: quick growing and hardy. Sow in good, rich soil, well drained and friable, a little every two weeks for summer varieties. Winter spinach must be sheltered. Water well. Buy varieties resistant to blight. Harvest late spring to autumn. Pick continuously from the outside. Steam or boil (in water left on leaves after washing); or make spinach soup.

***Further Reading***

***The Complete Urban Farmer*** by David Wickers (Julian Friedmann); ***Your Kitchen Garden*** by George Seddon (Mitchell Beazely); ***The Vegetable Garden Displayed*** (The Royal Horticultural Society); ***Food Crops From Your Garden or Allotment*** by Brian Turner (Pan).

THE POP UP FOOD BOOK

# LOW COST DIET

## BEANS

Beans are a marvellously economical way of feeding a large family without giving them meat all the time. There are at least 21 different shapes, sizes, colours and flavours to choose from. Black beans are large and shiny and, like red beans, have a smooth soft texture and look good mixed with colourful foods. Butter beans make good salads; fresh broad beans freeze well; chick peas are used for hummus, haricot beans for cassoulet.
This bean soup will feed eight people:

1 lb beans (any kind except soy)
1 ½ pints water
2 large onions, sliced
2 garlic cloves, crushed in salt
3oz chopped parsley
½ teaspoon dried thyme
1 tin tomato paste
juice of ½ lemon
salt and pepper

Bring water to the boil and add the beans. Cover and simmer gently for two hours or until nearly tender.
Add the rest of the ingredients and simmer for a further 30 minutes. You can substitute wine for half the water if you like.

Further reading:
***Diet for a small Planet*** by Frances Moore Lappe (Pan books);
***Whole Earth Cookbook*** by Sharon Cadwallader and Judi Ohr (Penguin 1973);
***The Bean Book*** by Rose Elliot

## BREAD

Soda breads are the easiest to make but don't stay fresh long. Bread is delicious made from wholemeal flour but a bit lighter if you put a proportion of white flour in.

Yeast bread is easy to make but does take time. You don't have to stay with it all the time. Bread left in a warm place will rise in an hour or under; left in the fridge it will take all night so organise your timetable as you will. For bread making the ingredients should all be warm.

Pitta is unusual but there are so many ways of eating it that it should be included in any bread maker's repertoire. It can be stuffed with meat or salad for school dinners or quick snacks or as the correct accompaniment to hummus or taramasalata before dinner.

This recipe will make six Pitta.

1 lb white flour
1 tsp salt
½ oz fresh yeast (or 1 tsp dried)
½ pint tepid water

Dissolve the yeast in the water. Put the flour and salt into a bowl, make a well in the centre and pour in the yeast mixture. Mix to a firm dough. Turn onto a floured surface and knead for ten minutes until dough is firm and elastic. Shape dough into a ball and put into a lightly greased bowl. Brush with oil. Cover with polythene and leave to rise until doubled in bulk. Put two greased baking sheets into the oven (450 F, 250 C, Gas 8). They must be hot, or the pitta won't rise. Knead the risen dough for three minutes. Divide into six, dust with flour and roll out to form a flat oval 8in x 5in. Put onto floured board, cover with polythene and leave 20 to 30 minutes. Transfer quickly to hot baking sheets; bake for about 10 min or until beginning to turn brown.

### *More information*

***The Wholefood Cookbook*** by George Seddon and Jackie Burrow (Mitchell Beazley);
***Daily Bread*** by Freda Murray (from P. Murray);
***Yeast Cookery,*** (WI Publications);
***Use Your Loaf*** by Ursel Norman (Fontana);
***English Bread and Yeast Cookery*** by Elizabeth David (Allen Lane).

# PEOPLE and IDEAS

IT is ideas, not armies, which change the world and at first we thought the Little Green Book should have an alphabetical list of key ideas. But the history of ideas is the history of the people, famous or obscure, who happened to have them, and promote them. The following in no way pretends to be a definitive list. Rather it is a selection of thumbnail sketches of a few people whose ideas have been, and remain, persistently relevant to the themes of this book.

---

**CARSON, Rachel**, 1907-1964, gets into the pantheon with one book, *Silent Spring,* published in 1962. An instant best-seller, it awoke the world to the poisoning of the environment and the disruption of the ecological balance by persistent pesticides. It marked the opening of a new era of ecological awareness. Dismissed by some patronising academics as emotional, she was a professional biologist in the US government wildlife service who saw what was happening. *Silent Spring* showed that the voice of individual protest can force bureaucracies to take notice. Earlier Rachel Carson had written another best seller, *The Sea Around Us.*

---

**GALBRAITH, John Kenneth**, born 1908, Harvard professor and subversive mandarin who turned the conventional wisdom of free market economics upside down and inside out. In a series of books — *The Affluent Society, The New Industrial State, Economics and the Public Purpose* — Galbraith was a destroyer of myths, a changer of ideas and a coiner of famous phrases. He saw private affluence amid public squalor, he identified and christened the technostructure, he proclaimed the end of consumer sovereignty as consumer wants were processed and reconditioned to serve the survival of the producer.

**HAECKEL, Ernst Heinrich** (1834-1919) was a German biologist who first used the term *ecology* for the study of the inter-relationship between organisms and their environment. It is a valuable concept which has been broadened (and some would say coarsened) to take in such themes as urban ecology and the idea of ecology as a political movement. Biologists frown at the expansion of their concept, but readers of the *Little Green Book* find the continual inter-relatedness of, for example, man and his environment a salutary reminder that no man (or woman) is an island. We tamper with the delicate equilibrium between living creatures (including ourselves) and their habitat, at our peril.

---

**HARDIN, Garrett**, professor of biology, University of California, Santa Barbara, USA. In 1968 he published an essay, "The Tragedy of the Commons", which is still the subject of lively debate. His neo-Malthusian approach to the population explosion leads him to the conclusion that "the freedom to breed will bring ruin to all". On common land each individual will try to keep as many sheep or cattle as possible. The logic of the process leads remorselessly to over-stocking. The tragedy of the commons is that individuals can only thrive by pursuing their own interests (which means increasing the herd when possible) at the expense of the collective good (limiting the total herd). The same process applies whether the commons are considered to be open pasture or whales or fossil fuel. In practice the commons were virtually ended by the enclosures. Hardin appears to be suggesting a similar process with his "living in a lifeboat" theory — exist frugally, stop immigration, stop foreign aid. The implications are shocking. Rather than being on a lifeboat, are we not on a large liner, in the luxury of the Captain's stateroom, while the masses are starving in steerage? On the other hand, is it not true that foreign aid has often damaged the recipient, and benefitted only the giver? Do we not in practice persistently misuse the

**continued over**

freedom of the commons in every way from the over-killing of whales to such projects as Concorde which give benefit to few at the expense of many? Whether or not one agrees with Hardin, he has at least succeeded in forcing people to confront the moral dilemma rather than rely on future technical solutions.

---

**ILLICH, Ivan** (b. 1926) is an Austrian-born Latin American ex-priest writer who in a series of closely-argued little books (published in England by Marion Boyars and Fontana) attacks many of the assumptions of the modern world. *Deschooling Society* sees the education system as the enemy of education and the guarantor of social inequality and exploitation. *Tolls for Conviviality* diagnoses the industrial system as a machine for human enslavement. His most damaging criticism in the environmental field is of the professionalisation of knowledge, precisely because the greater the expertise, the power and the status of a profession, the smaller the opportunity for the citizen to make decisions. "It makes people dependent on having their knowledge produced for them. It leads to a paralysis of the moral and political imagination. This cognitive disorder rests on the illusion that the knowledge of the individual citizen is of less value than the 'knowledge' of science. The former is the opinion of individuals. It is merely subjective and is excluded from policies. The latter is 'objective' — defined by science and promulgated by expert spokesmen. This objective knowledge is viewed as a commodity which can be refined, constantly improved, accumulated and fed into a process called 'decision-making'. This new mythology of governance by the manipulation of knowledge-stock inevitably erodes reliance on government by people. Over-confidence in 'better knowledge' becomes a self-fulfilling prophecy. People first cease to trust their own judgement and then want to be told the truth about what they know. Overconfidence in 'better decision-making' first hampers people's ability to decide for themselves and then undermines their belief that they can decide."

---

**JACOBS, Jane** is an American journalist whose book *The Death and Life of Great American Cities* was the first to recognise that the redevelopment of the city had killed the life of the street, its conviviality, its continual interest and stimulation, its educative and self-policing functions. The rebuilt city, she rightly complained, has "junked the basic function of the city street, and with it, necessarily, the freedom of the city."

---

**KROPOTKIN, Peter** (1842-1921) was a Russian geographer and anarchist whose *Mutual Aid* (last reprinted by Allan Lane) was an attempt to rescue Darwin's theory of natural selection from those who used it to defend predatory capitalism, by demonstrating that the survival of a species depended more on co-operation than on competition. *Fields, Factories and Workshops* (last reprinted by Allen & Unwin) argued firstly that there is a trend for manufacturing industry to decentralise throughout the world and that production for a local market is inevitable; secondly that in consequence of this Britain could no longer rely on importing cheap food in exchange for exports, but that our basic food needs could be met by intensive small-scale agriculture; thirdly that the dispersal of industry in combination with farming and horticulture is rational and desirable; and fourthly that we need an education that prepares us for a combination of manual and intellectual work.

**continued over**

# PEOPLE and IDEAS

**continued**

**LOVINS, Amory,** Wandering American Friends-of-the-Earth egg-head, whose historic article in *Foreign Affairs* (October 1976), "Energy Strategy: The Road Not Taken", helped change the face of discussion of energy policy in the industrial west. His subsequent book, *Soft Energy Paths* (Penguin 1977), remains a seminal primer for the continuing debate on energy choices — with its distinction between "hard" energy systems (entailing a growth in centralised supply, through technologies like nuclear electricity) and the "soft" approach (emphasising conservation and decentralised energy systems reliant on sun, wind, biomass etc).

---

**MORRIS, William** (1834-1896) was an English craftsman, poet and socialist, who used to be dismissed as a romantic dreamer, but was in fact a profound thinker. His *News from Nowhere* is a picture of post-industrial society, and as his biographer Paul Thompson remarks, he continually foreshadows our own preoccupations: "the destruction by the international economy, not just of ancient cultures, but of the natural resources and ecology of the earth itself; the crippling of local independence by spreading centralisation and bureaucracy, the stifling of the natural creativity and zest for learning of children by institutionalised schooling; the cramming of working people into barrack-like housing. Morris stands alone among major socialist thinkers in being as concerned with housework and the home as with work and the factory."

---

**ORWELL, George** (pen-name of Eric Blair, 1903-1950) was an English novelist and essayist who strove to rescue socialism from the bullies and

totalitarians. His prewar novel *Coming Up For Air* gives an uncannily accurate picture of the degradation of the English country town, its economy, its shops and its environment. Lower Binfield, with its ugly houses, plastic food and garbage-filled ponds, is all around us. *Animal Farm* was his fairy tale which subsumed the history of all Marxist revolutions, past and future, and *1984* was his projection into the future of trends he saw in the Cold War climate of 1948. It's only a few years away now.

---

**PROUDHON, Pierre-Joseph** (1809-1865) was a French anarchist. His book *What is Property?* gave the answer that "Property is Theft", but he also declared that "Property is Freedom". He was talking in the first instance of the absentee landlord and in the second of the peasant proprietor. For Proudhon was one of the few thinkers on the Left to extol the virtues of peasant self-sufficiency. "In my father's house we breakfasted on maize porridge; at mid-day we ate potatoes; in the evening bacon soup, and that every day of the week. And despite the economists who praise the English diet, we, with that vegetarian feeding were fat and strong. Do you know why? Because we breathed the air of our own fields and lived from the produce of our own cultivation."

---

**SCHUMACHER, E.F.** 1911-1977) was a man with an idea whose time had come: that giantism was destroying industrial societies, and that small was admirable, desirable, reasonable, right, human and beautiful. His book *Small is Beautiful* (1972) was sub-titled "A study of economics as if people mattered". It presented a system of ideas for human-scale fitness for purpose to cope with the disenchantments of the 1970s whether in developed or developing countries. The practical application of Schumacher's guidelines for appropriate small-scale technology was propagated in many countries of the world through the Intermediate

**continued over**

Technology Development Group which he founded in London in 1966. Wilton Corner, 10 Grenfell Road, Beaconsfield, Bucks. Tel: 04946-3080.

---

**THOREAU, Henry David** (1817-1862) was an American naturalist whose *Walden: or Life in the Woods* is, on the face of it, simply an account of his attempt to live self-sufficiently, but is actually a profound and witty attack on industrial civilisation. "The mass of men lead lives of quiet desperation" he concluded, but he was no escapist. Thoreau was the one person in America to speak up publicly for John Brown before his execution and his *On the Duty of Civil Disobedience* inspired both Tolstoy and Gandhi.

---

**TURNER, John F.C.** (b. 1927) is an English architect who spent years helping squatters in Latin America and returned to this country just when the stalemate of housing policy was leading thoughtful people to seek alternatives. From his books *Freedom to Build* (Collier-Macmillan) and *Housing by People* (Marion Boyars) we can distill three neglected principles of housing. First that the important thing about housing is not what it *is* but what it *does* in people's lives, secondly that deficiencies and imperfections in your housing are infinitely more tolerable if they are your responsibility than if they are somebody else's, and finally the principle of dweller control. "When dwellers control the major decisions and are free to make their own contribution to the design, construction or management of their housing, both the process and the environment produced stimulate individual and social well-being. When people have no control over, nor responsibility for key decisions in the housing process, on the other hand, dwelling environments may instead become a barrier to personal fulfillment and a burden on the economy."

**TYME, John**, Old Testament prophet, formerly a lecturer at Sheffield Polytechnic, whose direct action protests at motorway inquiries in 1975-78 dramatised powerful arguments developed by himself and other environmental bodies against the national road-building programme. Tyme's protests struck strong chords of popular dislike of the Department of Transport's high-handedness in promoting road-building — and contributed to reductions in the road programme itself and to reform of public inquiry procedures. Tyme's book, *Motorways and Democracy* (Macmillan 1979) spells it all out.

**WARD, Barbara**, born 1914, populariser extraordinary of the environmental message, polymath historian, economist, philosopher, crusader for Third World rights, passionately articulate writer of books, maker of speeches, organiser of pressure groups, world conferences (Stockholm 1972, Habitat Vancouver 1976), doyenne of the NGOs, author of key books of the 1970s — *Only One Earth* (1972), *The Home of Man* (1976), with another to come, *The Conserving Society* (1979) — in which she has expounded a sensitive political and scientific view of (her adjectives) our single, beautiful, fragile and vulnerable planet.

**But while I pondered all these things, and how men fight and lose the battle, and the thing that they fought for comes about in spite of their defeat, and when it comes turns out not to be what they meant, and other men have to fight for what they meant under another name — while I pondered all this, John Ball began to speak again.**
*William Morris, A Dream of John Ball*

# THE GREEN IDEA AND PRESSURE POLITICS

THE "GREEN PERSPECTIVE" is not simply a matter of looking after your own patch. Many of the most important initiatives affecting how we live result from political decisions at local, national and international levels.

So to influence events affecting the environment, it may be necessary to ally yourself with other like-minded individuals — by joining one or more of many bodies who are promoting and campaigning for environmentally sound policies.

Britain has an abundance of independent pressure groups and local societies which have come into being to do precisely this. There are those which are pressing, with different emphases, for more prudent use of land and resources — amongst them the **Town and Country Planning Association** (TCPA), **Friends of the Earth,** the **Councils for the Protection of Rural England** (CPRE) **and Wales,** the **Civic Trust** and the **Conservation Society.** These are more specialist pressure groups — **Transport 2000** and the **Midlands Motorway Action Group** in the field of transport, the **Ramblers' Association** for footpaths, the **Council for National Parks,** the **Royal Society for the Protection of Birds,** the **Society for the Promotion of Nature Conservation** and the **Inland Waterways Association,** to name but a few. Then there are bodies which own and acquire land for public benefit — the **National Trust** and the **Landmark Trust** for example.

And there are also more down-to-earth organisations like the **British Trust for Conservation Volunteers** and the **National Trust's Acorn Camps,** with their practical conservation projects for people all over the country.

All of these bodies are independent of government. Most support themselves through their memberships and by their fund-raising. In their different ways, they are relentlessly active — pressuring Ministers and civil servants, monitoring the actions of government agencies, promoting projects, submitting evidence and proposals to official working parties, lobbying Parliaments, organising demonstrations, influencing the news media. Their credibility with the government and the freshness of their ideas depends on the energy and concern of their members.

All this activity achieves results. The present ferment on national energy policy — should we go nuclear or not? — is largely due to the enterprise of groups like Friends of the Earth in publicising the issues. Similarly growing debates about conflicts between agricultural development and conservation of nature and landscape, the future of heavy lorries, the plight of rural communities and the decline of public transport reflect consistent pressure group activity. Much of the most important legislation affecting nature conservation and countryside protection, the Countryside Act (1968) the Conservation of Wild Creatures and Wild Plants Act (1975), the Protection of Birds Act (1914-1967), the Endangered Species Act (1976), Civic Amenities Act (1957), has been similarly shaped. Historically, the existence of National Parks, Green Belt, and even the Town and Country Planning Acts, (the very statutes on which coherent land use planning rests in the UK) owe much to the campaigning of bodies like the TCPA and CPRE.

There is a perpetual need for coordinated action. The **Committee for Environmental Conservation** plays some role in this in Britain — while at an international level, the **European Environmental Bureau** (EEB) is having a growing influence on the institutions of the European Economic Community. The Bureau represent more than 40 national environmental bodies in

**continued over**

# THE GREEN IDEA AND PRESSURE POLITICS

**continued**

the nine member countries of the Community, including several from Britain. It has an office in Brussels and is an increasingly articulate critic of the EEC's energy, transport and agricultural policies. It is also campaigning for the revision of the Euratom Treaty and the Treaty of Rome, in an attempt to turn the Community towards greener, less exclusively economic pastures in the long term.

None of this web of national or international activity would be possible without the interest and expertise of local bodies. The counties of England, Wales and Scotland are peppered with local amenity societies, environmental action groups and nature conservation groups — more than in any other country in Europe. In 1975, there were more than 1,200 amenity societies registered with the Civic Trust. They are forces to be reckoned with, as many local authorities now recognise. A good local group can have a remarkable effect on a community, promoting retention of what is historically significant, encouraging tree planting, affecting local transport policies, influencing the character of local industrial development, and promoting general discussion of the priorities the local authority should be pursuing.

Everyone who wishes can find a role in all this activity, locally or nationally. The green idea finds everyday expression through bodies like these. And if you are dissatisfied with their performances, well, their priorities can be changed by active participation. They are there for the joining.

Details of further organisations will be found in the *Environmental Directory*, published by the Civic Trust, 17 Carlton House Terrace, London SW1, £1.40 including postage.

# THE GREEN ALLIANCE

THE Green Alliance was formed in October 1978, as a result of a meeting of individuals most of whom were already active in independent organisations in the environmental field.

It is an Association with a limited membership. The supporting organisation *Friends of The Green Alliance* has an open membership. The aim of the *Alliance* is to ensure that the political priorities of the UK are determined within an ecological perspective.

The *Alliance* believes that present political orthodoxies create more problems than they try to solve. Pursuit of growth in GNP, centralisation of decision and power and the frantic quest for a technological 'fix', are all the enemies of a balanced, harmonious and ecological society.

The *Alliance* aims to reach new publics in new ways. This book is an example. It will also seek to influence decisions in Westminster, Whitehall, the City and in industry. It will not do so from the stand-point of any ideology but from the ecological perspective, which stands for the changes in direction needed now to make a humane and civilized future possible.

Enquiries to The Green Alliance, Francis House, Francis Street, London SW1.

The Little Green Book
was produced by **Vole**
for the Green Alliance.

First published in Great Britain 1979

Wildwood House Limited
1 Prince of Wales Passage
117 Hampstead Road
London NW1 3EE



ISBN 0 7045 0381 6

Printed and bound by The Guernsey Press Company, The Channel Islands